Revise
A2

Mathematics

Peter Sherran
& Janet Crawshaw

Contents

Specification lists

AQA Mathematics

MODULE	SPECIFICATION TOPIC	CHAPTER REFERENCE	STUDIED IN CLASS	REVISED	PRACTICE QUESTIONS
Core 3 (C3)	Algebra and functions	1.2			
	Trigonometry	1.4			
	Exponentials and logarithms	1.2			
	Differentiation	1.5			
	Integration	1.6			
	Numerical methods	1.7			
Core 4 (C4)	Algebra and functions	1.1, AS1.1			
	Coordinate geometry	1.3			
	Sequences and series	1.1			
	Trigonometry	1.4			
	Exponentials and logarithms	1.6			
	Differentiation and integration	1.5, 1.6			
	Vectors	1.8			
Statistics 2 (S2)	Discrete random variables	AS4.3			
	Poisson distribution	AS4.3			
	Continuous random variables	2.1			
	Estimation	2.3			
	Hypothesis testing	2.4, 2.5			
	χ^2 tests	2.5			
Mechanics 2 (M2)	Moments and centres of mass	AS3.3, 3.4, 3.5			
	Kinematics	3.2, 3.3			
	Newton's laws of motion	3.2			
	Application of differential equations	3.2			
	Uniform circular motion	3.7			
	Work, energy and power	3.9, 3.10			
	Vertical circular motion	3.8			
Decision 2 (D2)	Critical path analysis	4.3			
	Allocation	4.5			
	Dynamic programming	–			
	Network flows	4.2			
	Linear programming	4.4			
	Game theory	4.1			

Examination analysis

The Advanced Level Mathematics GCE consists of AS (50%) and A2 (50%). AS consists of C1 + C2 + one of M1, S1, D1. A2 consists of C3 + C4 + one of M1, S1, D1, M2, S2, D2 (following any dependency rules).

C3	A2	Scientific/graphics calculator	1 hr 30 min exam	$16\frac{2}{3}\%$ of A Level
C4	A2	Scientific/graphics calculator	1 hr 30 min exam	$16\frac{2}{3}\%$ of A Level
M1/M2	AS/A2	Scientific/graphics calculator	1 hr 30 min exam*	$16\frac{2}{3}\%$ of A Level
S1/S2	AS/A2	Scientific/graphics calculator	1 hr 30 min exam*	$16\frac{2}{3}\%$ of A Level
D1/D2	AS/A2	Scientific/graphics calculator	1 hr 30 min exam	$16\frac{2}{3}\%$ of A Level

* 1 hr 15 min with coursework

EDEXCEL Mathematics

MODULE	SPECIFICATION TOPIC	CHAPTER REFERENCE	STUDIED IN CLASS	REVISED	PRACTICE QUESTIONS
Core 3 (C3)	Algebra and functions	1.1, 1.2, AS1.1			
	Trigonometry	1.4			
	Exponentials and logarithms	1.2			
	Differentiation	1.5			
	Numerical methods	1.7			
Core 4 (C4)	Algebra and functions	1.1			
	Coordinate geometry	1.3			
	Sequences and series	1.1			
	Differentiation	1.5			
	Integration	1.6			
	Vectors	1.8			
Statistics 2 (S2)	Binomial and Poisson distributions	AS4.3			
	Continuous random variables	2.1			
	Continuous distributions	2.1, 2.2			
	Hypothesis tests	2.4			
Mechanics 2 (M2)	Kinematics	3.1, 3.2, 3.3			
	Centre of mass	3.5			
	Work and energy	3.9			
	Collisions	3.6			
	Statics of rigid bodies	3.4			
Decision 2 (D2)	Transportation problems	–			
	Allocation problems	4.5			
	The travelling salesman problem	4.2			
	Game theory	4.1			
	Dynamic programming	–			

Examination analysis

The Advanced Level Mathematics GCE consists of AS (50%) and A2 (50%). AS consists of C1 + C2 + one of M1, S1, D1. A2 consists of C3 + C4 + one of M1, S1, D1, M2, S2, D2 (following any dependency rules).

C3	A2	Scientific/graphics calculator	1 hr 30 min exam	$16\frac{2}{3}$% of A Level
C4	A2	Scientific/graphics calculator	1 hr 30 min exam	$16\frac{2}{3}$% of A Level
M1/M2	AS/A2	Scientific/graphics calculator	1 hr 30 min exam	$16\frac{2}{3}$% of A Level
S1/S2	AS/A2	Scientific/graphics calculator	1 hr 30 min exam	$16\frac{2}{3}$% of A Level
D1/D2	AS/A2	Scientific/graphics calculator	1 hr 30 min exam	$16\frac{2}{3}$% of A Level

OCR Mathematics

MODULE	SPECIFICATION TOPIC	CHAPTER REFERENCE	STUDIED IN CLASS	REVISED	PRACTICE QUESTIONS
Core 3 (C3)	Algebra and functions	1.2, 1.6			
	Trigonometry	1.4			
	Differentiation and integration	1.5, 1.6			
	Numerical methods	1.7			
Core 4 (C4)	Algebra and graphs	1.1, 1.3			
	Differentiation and integration	1.5, 1.6			
	First order differential equations	1.6			
	Vectors	1.8			
Statistics 2 (S2)	Continuous random variables	2.1			
	The normal distribution	AS4.5, 2.2			
	The Poisson distribution	AS4.3, 2.2			
	Sampling and hypothesis tests	2.3, 2.4			
Mechanics 2 (M2)	Centre of mass	3.5			
	Equilibrium of a rigid body	3.4			
	Motion of a projectile	3.1			
	Uniform motion in a circle	3.7			
	Coefficient of restitution; impulse	3.6			
	Energy, work and power	3.9			
Decision 2 (D2)	Game theory	4.1			
	Flows in a network	4.2			
	Matching and allocation problems	4.5			
	Critical path analysis	4.3			
	Dynamic programming	–			

Examination analysis

The Advanced Level Mathematics GCE consists of AS (50%) and A2 (50%). AS consists of C1 + C2 + one of M1, S1, D1. A2 consists of C3 + C4 + one of M1, S1, D1, M2, S2, D2 (following any dependency rules).

C3	A2	Scientific/graphics calculator	1 hr 30 min exam	$16\frac{2}{3}$% of A Level
C4	A2	Scientific/graphics calculator	1 hr 30 min exam	$16\frac{2}{3}$% of A Level
M1/M2	AS/A2	Scientific/graphics calculator	1 hr 30 min exam	$16\frac{2}{3}$% of A Level
S1/S2	AS/A2	Scientific/graphics calculator	1 hr 30 min exam	$16\frac{2}{3}$% of A Level
D1/D2	AS/A2	Scientific/graphics calculator	1 hr 30 min exam	$16\frac{2}{3}$% of A Level

WJEC Mathematics

MODULE	SPECIFICATION TOPIC	CHAPTER REFERENCE	STUDIED IN CLASS	REVISED	PRACTICE QUESTIONS
Core 3 (C3)	Trigonometry	1.4			
	Functions	1.2			
	Exponentials and logarithms	1.2			
	Differentiation	1.5			
	Integration	1.6			
	Numerical methods	1.7			
Core 4 (C4)	Algebra and series	1.1			
	Trigonometry	1.4			
	Coordinate geometry	1.3			
	Differential equations	1.6			
	Integration	1.6			
	Vectors	1.8			
Statistics 2 (S2)	Uniform distribution	2.1			
	Normal distribution	AS4.5, 2.2			
	Random variables	AS4.3, 2.1			
	Distribution of sample mean	2.3			
	Hypothesis testing	2.4			
	Confidence intervals	2.3			
Mechanics 2 (M2)	Rectilinear motion	3.2			
	Dynamics of a particle	3.9, 3.10			
	Motion under gravity in 2-d	3.1			
	Vectors	3.3			
	Circular motion	3.7, 3.8			

Examination analysis

The Advanced Level Mathematics GCE consists of AS (50%) and A2 (50%). AS consists of C1 + C2 + one of M1, S1. A2 consists of C3 + C4 + one of M1, S1, M2, S2 (following any dependency rules).

C3	A2	Scientific/graphics calculator	1 hr 30 min exam	$16\frac{2}{3}$% of A Level
C4	A2	Scientific/graphics calculator	1 hr 30 min exam	$16\frac{2}{3}$% of A Level
M1/M2	AS/A2	Scientific/graphics calculator	1 hr 30 min exam	$16\frac{2}{3}$% of A Level
S1/S2	AS/A2	Scientific/graphics calculator	1 hr 30 min exam	$16\frac{2}{3}$% of A Level

CCEA Mathematics

MODULE	SPECIFICATION TOPIC	CHAPTER REFERENCE	STUDIED IN CLASS	REVISED	PRACTICE QUESTIONS
Core 3 (C3)	Algebra and functions	1.1, 1.2			
	Coordinate geometry	1.3			
	Series	1.1			
	Trigonometry	1.4			
	Exponentials and logarithms	1.2, 1.6			
	Differentiation and integration	1.5, 1.6			
	Numerical methods	1.7			
Core 4 (C4)	Functions	1.2			
	Trigonometry	1.4			
	Differentiation	1.5			
	Integration	1.6			
	Vectors	1.8			
Statistics 2 (S2)	Expectation algebra	2.1			
	Estimation and sampling	2.3			
	Hypothesis testing	2.4, 2.5			
	Bivariate distributions	2.5, AS4.6			
Mechanics 2 (M2)	Displacement, velocity, acceleration	3.2, 3.3			
	Variable acceleration	3.2, 3.3			
	Projectiles	3.1			
	Circular motion	3.7, 3.8			
	Work, energy and power	3.9			

Examination analysis

The Advanced Level Mathematics GCE consists of AS (50%) and A2 (50%). AS consists of C1 + C2 + one of M1, S1. A2 consists of C3 + C4 + one of M1, S1, M2, S2 (following any dependency rules).

C3	A2	Scientific/graphics calculator	1 hr 30 min exam	$16\frac{2}{3}$% of A Level
C4	A2	Scientific/graphics calculator	1 hr 30 min exam	$16\frac{2}{3}$% of A Level
M1/M2	AS/A2	Scientific/graphics calculator	1 hr 30 min exam	$16\frac{2}{3}$% of A Level
S1/S2	AS/A2	Scientific/graphics calculator	1 hr 30 min exam	$16\frac{2}{3}$% of A Level

AS/A2 Level Mathematics courses

AS and A2

All Mathematics A Level courses are in two parts, with three separate modules in each part. Students first study the AS (Advanced Subsidiary) course. Some will then go on to study the second part of the A Level course, called A2. Advanced Subsidiary is assessed at the standard expected halfway through an A Level course: i.e., between GCSE and Advanced GCE. This means that the AS and A2 courses are designed so that difficulty steadily increases:

- AS Mathematics builds from GCSE Mathematics
- A2 Mathematics builds from AS Mathematics.

How will you be tested?

Assessment units

For AS Mathematics, you will be tested by three assessment units. For the full A Level in Mathematics, you will take a further three units. AS Mathematics forms 50% of the assessment weighting and A2 Mathematics forms the other 50% for the full A Level.

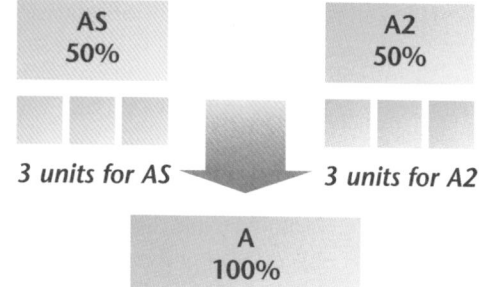

Most units can be taken in either January or June, but some A2 units must be taken at the end of the course. There is a lot of flexibility about when exams can be taken and the diagram below shows just some of the ways that the assessment units may be taken for AS and A Level Mathematics.

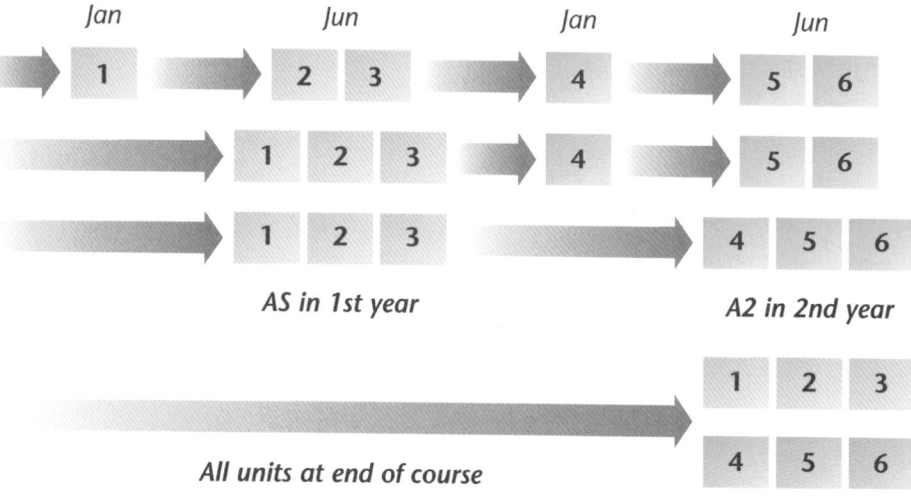

If you are disappointed with a module result, you can resit the module. There is no restriction on the number of times a module may be attempted. The best available results for each module will count towards the final grade.

A2 progression

After having studied AS Mathematics, to continue studying Mathematics to A Level you will need to take three further units of Mathematics, at least two of which are at A2 standard.

Synoptic assessment

Synoptic assessment in Mathematics addresses candidates' understanding of the connections between different elements of the subject. It involves the explicit drawing together of knowledge, understanding and skills learned in different parts of the A Level course, through using and applying methods developed earlier to solving problems. Making and understanding connections in this way is intrinsic to learning mathematics.

For A2 modules you will be asked to demonstrate knowledge of modules studied at AS Level. The contents of C1 and C2 are assumed for C3 and the contents of C1, C2 and C3 are assumed for C4. Similarly in the application modules, a knowledge of S1 is assumed for S2, a knowledge of M1 is assumed for M2 and a knowledge of D1 is assumed for D2.

Coursework

Coursework may form part of your A Level Mathematics course, depending on which specification you study.

Key skills

It is important that you develop your key skills throughout your AS and A2 courses. These are important skills that you need whatever you do beyond AS and A Levels. To gain the key skills qualification, which is equivalent to an AS Level, you will need to collect evidence together in a 'portfolio' to show that you have attained Level 3 in Communication, Application of number and Information technology. You will also need to take a formal test in each key skill. You will have many opportunities during AS and A2 Mathematics to develop your key skills.

It is a worthwhile qualification, as it demonstrates your ability to put your ideas across to other people, collect data and use up-to-date technology in your work.

What skills will I need?

For A Level Mathematics (AS and A2), you will be tested by assessment objectives: these are the skills and abilities that you should have acquired by studying the course. The assessment objectives for A Level Mathematics are shown below.

Candidates should be able to:

- recall, select and use their knowledge of mathematical facts, concepts and techniques in a variety of contexts

- construct rigorous mathematical arguments and proofs through use of precise statements, logical deduction and inference and by the manipulation of mathematical expressions, including the construction of extended arguments for handling substantial problems presented in unstructured form

- recall, select and use their knowledge of standard mathematical models to represent situations in the real world; recognise and understand given representations involving standard models; present and interpret results from such models in terms of the original situation, including discussion of the assumptions made and refinement of such models

- comprehend translations of common realistic contexts into mathematics; use the results of calculations to make predictions, or comment on the context; and, where appropriate, read critically and comprehend longer mathematical arguments or examples of applications

- use contemporary calculator technology and other permitted resources (such as formulae booklets or statistical tables) accurately and efficiently; understand when not to use such technology, and its limitations; give answers to appropriate accuracy.

Exam technique

What are the examiners looking for?

Examiners use certain words in their instructions to let you know what they are expecting in your answer. Make sure that you know what they mean so that you can give the right response.

Write down, state

You can write your answer without having to show how it was obtained. There is nothing to prevent you doing some working if it helps you, but if you are doing a lot then you might have missed the point.

Calculate, find, determine, show, solve

Make sure that you show enough working to justify the final answer or conclusion. Marks will be available for showing a correct method.

Deduce, hence

This means that you are expected to use the given result to establish something new. You must show all of the steps in your working.

Draw

This is used to tell you to plot an accurate graph using graph paper. Take note of any instructions about the scale that must be used. You may need to read values from your graph.

Sketch

If the instruction is to sketch a graph then you don't need to plot the points but you will be expected to show its general shape and its relationship with the axes. Indicate the positions of any turning points and take particular care with any asymptotes.

Find the exact value

This instruction is usually given when the final answer involves an irrational value such as a logarithm, e, π or a surd. You will need to demonstrate that you can manipulate these quantities so don't just key everything into your calculator or you will lose marks.

If a question requires the final answer to be given to a specific level of accuracy then make sure that you do this or you might needlessly lose marks.

Some dos and don'ts

Dos

Do read the question

- Make sure that you are clear about what you are expected to do. Look for some structure in the question that may help you take the right approach.
- Read the question *again* after you have answered it as a quick check that your answer is in the expected form.

Do use diagrams

- In some questions, particularly in mechanics, a clearly labelled diagram is essential. Use a diagram whenever it may help you understand or represent the problem that you are trying to solve.

Do take care with notation

- Write clearly and use the notation accurately. Use brackets when they are required.
- Even if your final answer is wrong, you may earn some marks for a correct expression in your working.

Do learn relevant formulae

- Each module specification contains a list of formulae that *will not be given* in the examination. Make sure that you learn these formulae and practise using them.
- Some formulae *will be given* in the Examination Formulae Booklet. Make sure you know what they are and where to find them. This will help you refer to them quickly in the examination so that you don't waste time.

Do avoid silly answers

- Check that your final answer is sensible within the context of the question.

Do make good use of time

- Choose the order in which you answer the questions carefully. Do the ones that are easiest for you first.
- Set yourself a time limit for a question depending on the number of marks available.
- Be prepared to leave a difficult part of a question and return to it later if there is time.
- Towards the end of the exam make sure that you pick up all of the easy marks in any questions that you haven't got time to answer fully.

Don'ts

Don't work with rounded values

- There may be several stages in a solution that produce numerical values. Rounding errors from earlier stages may distort your final answer. One way to avoid this is to make use of your calculator memories to store values that you will need again.

Don't cross out work that may be partly correct

- It's tempting to cross out something that hasn't worked out as it should. Avoid this unless you have time to replace it with something better.

Don't write out the question

- This wastes time. The marks are for your solution!

What grade do you want?

Everyone should be able to improve their grades but you will only manage this with a lot of hard work and determination. The details given below describe a level of performance typical of candidates achieving grades A, C or E. You should find it useful to read and compare the expectations for the different levels and to give some thought to the areas where you need to improve most.

Grade A candidates

- Recall or recognise almost all the mathematical facts, concepts and techniques that are needed, and select appropriate ones to use in a variety on contexts.
- Manipulate mathematical expressions and use graphs, sketches and diagrams, all with high accuracy and skill.
- Use mathematical language correctly and proceed logically and rigorously through extended arguments or proofs.
- When confronted with unstructured problems they can often devise and implement an effective solution strategy.
- If errors are made in their calculations or logic, these are sometimes noticed and corrected.
- Recall or recognise almost all the standard models that are needed, and select appropriate ones to represent a wide variety of situations in the real world.
- Correctly refer results from calculations using the model to the original situation; they give sensible interpretations of their results in the context of the original realistic situation.
- Make intelligent comments on the modelling assumptions and possible refinements to the model.
- Comprehend or understand the meaning of almost all translations into mathematics of common realistic contexts.
- Correctly refer the results of calculations back to given context and usually make sensible comments or predictions.
- Can distil the essential mathematical information from extended pieces of prose having mathematical content.
- Comment meaningfully on the mathematical information.
- Make appropriate and efficient use of contemporary calculator technology and other permitted resources, and are aware of any limitations to their use.
- Present results to an appropriate degree of accuracy.

Grade C candidates

- Recall or recognise most of the mathematical facts, concepts and techniques that are needed, and usually select appropriate ones to use in a variety of contexts.
- Manipulate mathematical expressions and use graphs, sketches and diagrams, all with a reasonable level of accuracy and skill.
- Use mathematical language with some skill and sometimes proceed logically through extended arguments or proofs.
- When confronted with unstructured problems they sometimes devise and implement an effective and efficient solution strategy.

- Occasionally notice and correct errors in their calculations.
- Recall or recognise most of the standard models that are needed and usually select appropriate ones to represent a variety of situations in the real world.
- Often correctly refer results from calculations using the model to the original situation, they sometimes give sensible interpretations of their results in context of the original realistic situation.
- Sometimes make intelligent comments on the modelling assumptions and possible refinements to the model.
- Comprehend or understand the meaning of most translations into mathematics of common realistic contexts.
- Often correctly refer the results of calculations back to the given context and sometimes make sensible comments or predictions.
- Distil much of the essential mathematical information from extended pieces of prose having mathematical content.
- Give some useful comments on this mathematical information.
- Usually make appropriate and effective use of contemporary calculator technology and other permitted resources, and are sometimes aware of any limitations to their use.
- Usually present results to an appropriate degree of accuracy.

Grade E candidates

- Recall or recognise some of the mathematical facts, concepts and techniques that are needed, and sometimes select appropriate ones to represent to use in some contexts.
- Manipulate mathematical expressions and use graphs, sketches and diagrams, all with some accuracy and skill.
- Sometimes use mathematical language correctly and occasionally proceed logically through extended arguments or proofs.
- Recall or recognise some of the standard models that are needed and sometimes select appropriate ones to represent a variety of situations in the real world.
- Sometimes correctly refer results from calculations using the model to the original situation; they try to interpret their results in the context of the original realistic situation.
- Sometimes comprehend or understand the meaning of translations in mathematics of common realistic contexts.
- Sometimes correctly refer the results of calculations back to the given context and attempt to give comments or predictions.
- Distil some of the essential mathematical information from extended pieces of prose having mathematical content; they attempt to comment on this mathematical information.
- Candidates often make appropriate and efficient use of contemporary calculator technology and other permitted resources.
- Often present results to an appropriate degree of accuracy.

The table below shows how your uniform standardised mark is translated.

USM	80%	70%	60%	50%	40%
grade	A	B	C	D	E

Four steps to successful revision

Step 1: Understand

- Study the topic to be learned slowly. Make sure you understand the logic or important concepts.
- Mark up the text if necessary – underline, highlight and make notes.
- Re-read each paragraph slowly.

GO TO STEP 2

Step 2: Summarise

- Now make your own revision note summary:
  What is the main idea, theme or concept to be learned?
  What are the main points? How does the logic develop?
  Ask questions: Why? How? What next?
- Use bullet points, mind maps, patterned notes.
- Link ideas with mnemonics, mind maps, crazy stories.
- Note the title and date of the revision notes
  (e.g. Mathematics: Differentiation, 3rd March).
- Organise your notes carefully and keep them in a file.

This is now in **short-term memory**. You will forget 80% of it if you do not go to Step 3.
GO TO STEP 3, but first take a 10 minute break.

Step 3: Memorise

- Take 25 minute learning 'bites' with 5 minute breaks.
- After each 5 minute break test yourself:
  Cover the original revision note summary
  Write down the main points
  Speak out loud (record on tape)
  Tell someone else
  Repeat many times.

The material is well on its way to **long-term memory**.
You will forget 40% if you do not do step 4. **GO TO STEP 4**

Step 4: Track/Review

- Create a Revision Diary (one A4 page per day).
- Make a revision plan for the topic, e.g. 1 day later, 1 week later, 1 month later.
- Record your revision in your Revision Diary, e.g.
  Mathematics: Differentiation, 3rd March 25 minutes
  Mathematics: Differentiation, 5th March 15 minutes
  Mathematics: Differentiation, 3rd April 15 minutes
  ... and then at monthly intervals.

Chapter 1

Core 3 and Core 4 (Pure Mathematics)

The following topics are covered in this chapter:

- Algebra and series
- Functions
- Coordinate geometry
- Trigonometry

- Differentiation
- Integration
- Numerical methods
- Vectors

1.1 Algebra and series

After studying this section you should be able to:

- simplify rational expressions
- express rational expressions in partial fractions
- use the binomial expansion $(1 + x)^n$ for any rational n

LEARNING SUMMARY

Rational expressions

AQA	C4
EDEXCEL	C3
OCR	C4
WJEC	C4
CCEA	C3

A **rational expression** is of the form $\dfrac{f(x)}{g(x)}$, where $f(x)$ and $g(x)$ are polynomials in x.

To simplify a rational expression, factorise both $f(x)$ and $g(x)$ as far as possible, then cancel any factors that appear in both the numerator and denominator.

Examples

(a) $\dfrac{2x^2 + 6x}{2x^2 + 7x + 3} = \dfrac{2x(x + 3)}{(2x + 1)(x + 3)}$

$$= \dfrac{2x}{2x + 1}$$

> This cannot be cancelled any further.

> Factorise $x^2 - 4$ using difference between two squares.

(b) $\dfrac{3x^2 - 12}{2 - x} = \dfrac{3(x^2 - 4)}{2 - x}$

> $x - 2 = -(2 - x)$.

$$= \dfrac{3(x - 2)(x + 2)}{(2 - x)}$$

$$= -3(x + 2)$$

When adding and subtracting algebra fractions, always use the lowest common denominator, for **example**,

> First factorise the denominators.

> The lowest common multiple of the denominators is $(x + 3)(x - 4)(x - 3)$.

$$\dfrac{x - 2}{x^2 - x - 12} + \dfrac{4}{x^2 - 9} = \dfrac{x - 2}{(x + 3)(x - 4)} + \dfrac{4}{(x + 3)(x - 3)}$$

$$= \dfrac{(x - 3)(x - 2) + 4(x - 4)}{(x + 3)(x - 4)(x - 3)}$$

$$= \dfrac{x^2 - 5x + 6 + 4x - 16}{(x + 3)(x - 4)(x - 3)}$$

$$= \dfrac{x^2 - x - 10}{(x + 3)(x - 4)(x - 3)}$$

Key points from AS

Remainder theorem page 22
Algebraic division page 21

Partial fractions

AQA	C4
EDEXCEL	C4
OCR	C4
WJEC	C4
CCEA	C3

It is sometimes possible to decompose a rational function into **partial fractions**. This format is often useful when differentiating, integrating or expanding functions as a series. Look out for the following types of partial fractions.

Denominator contains linear factors only

$$\frac{x+7}{(x-2)(x+1)} \equiv \frac{A}{x-2} + \frac{B}{x+1} = \frac{A(x+1)+B(x-2)}{(x-2)(x+1)}$$

$$\Rightarrow x+7 \equiv A(x+1) + B(x-2)$$

> The symbol $\equiv$ indicates that this is an identity. It is true for all values of x.

> These substitutions are chosen so that each factor in turn becomes zero.

Let $x = -1$, then $6 = -3B \Rightarrow B = -2$
Let $x = 2$, then $9 = 3A \Rightarrow A = 3$

> Examination questions could have up to three linear factors in the denominator.

$$\therefore \frac{x+7}{(x-2)(x+1)} \equiv \frac{3}{x-2} - \frac{2}{x+1}$$

Denominator contains repeated linear factors

> Do not use $(x+3)(x+1)(x+1)^2$ as the common denominator as it is not the lowest common multiple.

$$\frac{6x^2+15x+7}{(x+3)(x+1)^2} \equiv \frac{A}{x+3} + \frac{B}{x+1} + \frac{C}{(x+1)^2}$$

$$\equiv \frac{A(x+1)^2 + B(x+3)(x+1) + C(x+3)}{(x+3)(x+1)^2}$$

$$\Rightarrow 6x^2+15x+7 \equiv A(x+1)^2 + B(x+3)(x+1) + C(x+3)$$

> Use a mixture of substitution and equating coefficients to find A, B and C.

Let $x = -1$, then $-2 = 2C \Rightarrow C = -1$
Let $x = -3$, then $16 = 4A \Rightarrow A = 4$
Equate x^2 terms: $6 = A + B$ and since $A = 4$, $B = 2$.

$$\therefore \frac{6x^2+15x+7}{(x+3)(x+1)^2} \equiv \frac{4}{x+3} + \frac{2}{x+1} - \frac{1}{(x+1)^2}$$

> The degree of a polynomial is its highest power of x, for example $x^3 + 2x^2 - 3x + 4$ has degree 3.

Only **proper fractions** (when the numerator is of lower degree than the denominator) can be written in partial fraction form. For **improper fractions**, divide the denominator into the numerator first. This can be done by long division or by using identities.

Binomial expansion for rational n

AQA	C4
EDEXCEL	C4
OCR	C4
WJEC	C4
CCEA	C3

> **KEY POINT**
>
> When n is rational, $(1+x)^n$ can be written as a series, using the **binomial expansion**
>
> $$(1+x)^n = 1 + nx + \frac{n(n-1)}{2!}x^2 + \frac{n(n-1)(n-2)}{3!}x^3 + \dots$$

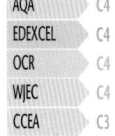

Key points from AS

- **Binomial expansion when n is a positive integer**
  Revise AS page 48

If n is a positive integer, the series terminates at the term in x^n.

For all other values of n, the series is infinite and converges provided that $|x| < 1$.

Example

(a) Expand $\sqrt{1-4x}$ as an ascending series in x, as far as the term in x^3, giving the set of values of x for which the series is valid.

(b) By substituting $x = 0.01$ into the series expansion of $\sqrt{1-4x}$, find $\sqrt{96}$ correct to 4 decimal places.

> $1 - 4x = 1 + (-4x)$, so use $(1 + x)^n$ with $n = \frac{1}{2}$ and substituting $(-4x)$ for x.

(a) $\sqrt{1-4x} = (1-4x)^{\frac{1}{2}}$

$$= 1 + \tfrac{1}{2}(-4x) + \frac{(\frac{1}{2})(-\frac{1}{2})}{2!}(-4x)^2 + \frac{(\frac{1}{2})(-\frac{1}{2})(-\frac{3}{2})}{3!}(-4x)^3 + \ldots$$

$$= 1 - 2x - 2x^2 - 4x^3 + \ldots$$

> You should also state the set of values of x for which the expansion is valid, even when it is not specifically requested in the question.

The series is valid provided that $|4x| < 1$, i.e. $|x| < \frac{1}{4}$.

(b) Substituting $x = 0.01$ into $\sqrt{1-4x}$ gives $\sqrt{0.96} = \sqrt{\dfrac{96}{100}} = \dfrac{\sqrt{96}}{10}$.

So $\sqrt{96} = 10(1 - 2(0.01) - 2(0.01)^2 - 4(0.01)^3 + \ldots)$

$\qquad\quad = 10(1 - 0.02 - 0.0002 - 0.000004 +)$

$\qquad\quad = 9.7980 \ (4 \ \text{d.p.})$

The function to be expanded must be in the form $(1 + \ldots)^n$ or $(1 - \ldots)^n$. You may have to do some careful algebraic manipulation to get it in this form.

Example

Give the first four terms in the series expansion of $\dfrac{1}{(2+x)}$.

$$\frac{1}{(2+x)} = \frac{1}{2\left(1 + \dfrac{x}{2}\right)}$$

$$= \frac{1}{2}\left(1 + \frac{x}{2}\right)^{-1}$$

> Alternatively,
> $(2+x)^{-1} = 2^{-1}\left(1 + \dfrac{x}{2}\right)^{-1}$.

$$= \frac{1}{2}\left(1 + (-1)\left(\frac{x}{2}\right) + \frac{(-1)(-2)}{2!}\left(\frac{x}{2}\right)^2 + \frac{(-1)(-2)(-3)}{3!}\left(\frac{x}{2}\right)^3 + \ldots\right)$$

$$= \frac{1}{2}\left(1 - \frac{x}{2} + \frac{x^2}{4} - \frac{x^3}{8} + \ldots\right)$$

$$= \frac{1}{2} - \frac{x}{4} + \frac{x^2}{8} - \frac{x^3}{16} + \ldots$$

Restriction on x: $\left|\dfrac{x}{2}\right| < 1$, i.e. $|x| < 2$.

Progress check

1 Simplify:

(a) $\dfrac{2x^2 + 3x}{2x^2 + x - 3}$ (b) $\dfrac{9x - x^3}{3x^2 + x^3}$

2 Express as a single fraction in its simplest form:

(a) $\dfrac{3(x-4)}{(x+2)(x-1)} + \dfrac{2(x+1)}{x-1}$

(b) $\dfrac{x^2 - 3x + 2}{x^2 + 3x - 4} - \dfrac{1}{(x+4)^2}$

3 $\dfrac{4x^2 + x + 1}{(x+2)(x-3)(x+1)} \equiv \dfrac{A}{x+2} + \dfrac{B}{x-3} + \dfrac{C}{x+1}$

Find the values of A, B and C.

4 Express in partial fractions:

$\dfrac{3x^2 + 4x + 5}{(x-1)^2(x+2)}$

5 Write as a series in ascending powers of x as far as the term in x^3, stating the values of x for which the expansion is valid:

(a) $\dfrac{1}{1+x}$ (b) $(1 - 3x)^{-2}$ (c) $\sqrt{4+x}$

5 (a) $1 - x + x^2 - x^3 + \cdots$ for $|x| < 1$
(b) $1 + 6x + 27x^2 + 108x^3 + \cdots$ for $|x| < \frac{1}{3}$
(c) $2 + \frac{1}{4}x - \frac{1}{64}x^2 + \frac{3}{512}x^3 + \cdots$ for $|x| < 4$

4 $\dfrac{2}{x-1} + \dfrac{4}{(x-1)^2} + \dfrac{1}{x+2}$

3 $A = 3$, $B = 2$, $C = -1$

2 (a) $\dfrac{2x^2 + 9x - 8}{(x+2)(x-1)}$ (b) $\dfrac{x^2 + 2x - 9}{(x+4)^2}$

1 (a) $\dfrac{x}{x-1}$ (b) $\dfrac{3-x}{x}$

1.2 Functions including exponential and logarithmic

After studying this section you should be able to:

- *use function notation and understand domain and range*
- *understand composite functions and find the inverse of a one-one function*
- *sketch the graph of the inverse of a function from the graph of the function*
- *understand the properties of the modulus function*
- *use transformations to sketch graphs of related functions*
- *understand the definitions of e^x and $\ln x$*

LEARNING SUMMARY

Functions

AQA	C3
EDEXCEL	C3
OCR	C3
WJEC	C3
CCEA	C4

x maps to y
y is the image of x.

$f(x)$ is read as 'f of x'.

A **function** may be thought of as a rule which takes each member x of a set and assigns, or **maps**, it to some value y known as its **image**.

$$x \longrightarrow \boxed{\text{Function}} \longrightarrow y$$

A letter such as f, g or h is often used to stand for a function. The function which squares a number and adds on 5, for example, can be written as $f(x) = x^2 + 5$. The same notation may also be used to show how a function affects particular values. For this function, $f(4) = 4^2 + 5 = 21$, $f(-10) = (-10)^2 + 5 = 105$ and so on.

An alternative notation for the same function is $f: x \mapsto x^2 + 5$.

> The set of values on which the function acts is called the **domain** and the corresponding set of image values is called the **range**.
>
> **KEY POINT**

The function $f(x) = x + 2$ with domain $\{3, 6, 10\}$ is shown below.

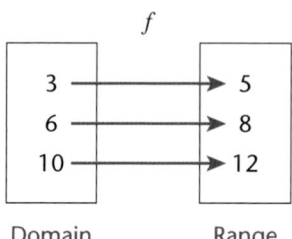

Domain   Range

The symbol $\mathbb{R}$ is often used to stand for the complete set of numbers on the number line. This is the set of Real numbers.

The domain of a function may be an infinite set such as $\mathbb{R}$ but, in some cases, particular values must be omitted for the function to be valid.

For example, the function $f(x) = \dfrac{x + 2}{x - 3}$ cannot have 3 in its domain since division by zero is undefined.

Under a function, every member of the domain must have only one image. But, it is possible for different values in the domain to have the same image.

There are different types of function.

> If for each element y in the range, there is *a unique* value of x such that $f(x) = y$ then f is a **one–one** function. If for any element y in the range, there is more than one value of x satisfying $f(x) = y$ then f is **many–one**.
>
> **KEY POINT**

Examples

These examples show the importance of including the domain of the function in its definition.

The function $f(x) = 2x$ with domain $\mathbb{R}$ is one–one.

The function $f(x) = x^2$ with domain $\mathbb{R}$ is many–one.

The function $f(x) = x^2$ with domain $x > 0$ is one–one.

The function $f(x) = \sin x$ with domain $0° \leqslant x \leqslant 360°$ is many–one.

The function $f(x) = \sin x$ with domain $-90° \leqslant x \leqslant 90°$ is one–one.

Composition of functions

AQA   C3
EDEXCEL   C3
OCR   C3
WJEC   C3
NICCEA   C4

The diagram shows how two functions f and g may be combined.

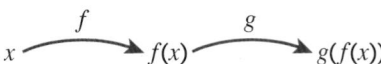

In this case, f is applied first to some value x giving $f(x)$. Then g is applied to the value $f(x)$ to give $g(f(x))$. This is usually written as $gf(x)$ and gf can be thought of as a new **composite function** defined from the functions f and g.

For example, if $f(x) = 3x$, $x \in \mathbb{R}$ and $g(x) = x + 2$, $x \in \mathbb{R}$ then
$$gf(x) = g(3x) = 3x + 2.$$

The order in which the functions are applied is important. The composite function fg is found by applying g first and then f.

In this case, $fg(x) = f(x + 2) = 3x + 6$.

Generally speaking, when two functions f and g are defined, the composite functions fg and gf will not be the same.

The inverse of a function

AQA   C3
EDEXCEL   C3
OCR   C3
WJEC   C3
NICCEA   C4

The **inverse of a function** f is a function, usually written as f^{-1}, that *undoes* the effect of f. So the inverse of a function which adds 2 to every value, for example, will be a function that subtracts 2 from every value.

This can be written as $f(x) = x + 2$, $x \in \mathbb{R}$ and $f^{-1}(x) = x - 2$, $x \in \mathbb{R}$.

The domain of f^{-1} is given by the range of f.

Notice that $f^{-1}f(x) = f^{-1}(x + 2) = x$ and that $ff^{-1}(x) = f(x - 2) = x$.

A function can be either one–one or many–one, but only functions that are one–one can have an inverse. The reason is, that reversing a many–one function would give a mapping that is one–many, and this cannot be a function.

The diagram shows a many–one function. It does not have an inverse.

Every value in the domain of a function can have only one image. This is an important property of functions.

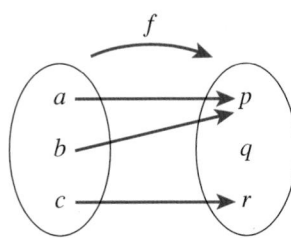

If you reverse this diagram then p would have more than one image.

Some important examples of restricting the domain of a function so that it will have an inverse are found in trigonometry. See page 32.

You can turn a many–one function into a one–one function by restricting its domain.

For example, the function $f(x) = x^2$, $x \in \mathbb{R}$ is many–one and so it does not have an inverse. However, the function $f(x) = x^2$, $x > 0$ is one–one and $f^{-1}(x) = \sqrt{x}$, $x > 0$.

Finding the inverse of a one–one function

AQA — C3
EDEXCEL — C3
OCR — C3
WJEC — C3
CCEA — C4

You can turn $f^{-1}(y)$ into $f^{-1}(x)$ by replacing every y with an x.

One way to find the inverse of a one–one function f is to write $y = f(x)$ and then rearrange this to make x the subject so that $x = f^{-1}(y)$. The inverse function is then usually defined in terms of x to give $f^{-1}(x)$. The domain of f^{-1} is the range of f.

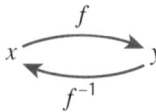

Example   Find the inverse of the function $f(x) = \dfrac{x+2}{x-3}$, $x \neq 3$.

Start by writing $y = f(x)$.

Define $y = \dfrac{x+2}{x-3}$

Rearrange to find x.

then $y(x-3) = x+2$

Multiply out the brackets.

$xy - 3y = x + 2$

Collect the x terms on one side of the equation.

$xy - x = 3y + 2$

$x(y - 1) = 3y + 2$

$x = \dfrac{3y+2}{y-1}$.

$x = f^{-1}(y)$ so this defines the inverse function in terms of y.

It is usual to express the inverse function in terms of x.

So $f^{-1}(x) = \dfrac{3x+2}{x-1}$, $x \neq 1$.

The denominator cannot be allowed to be zero so 1 is not in the domain of f^{-1}.

Here are the graphs of $y = \dfrac{x+2}{x-3}$ and its inverse $y = \dfrac{3x+2}{x-1}$

The graph of the inverse function is the reflection of the original graph in the line $y = x$. This is true for any inverse function but the same scale must be used on both axes or the effect is distorted.

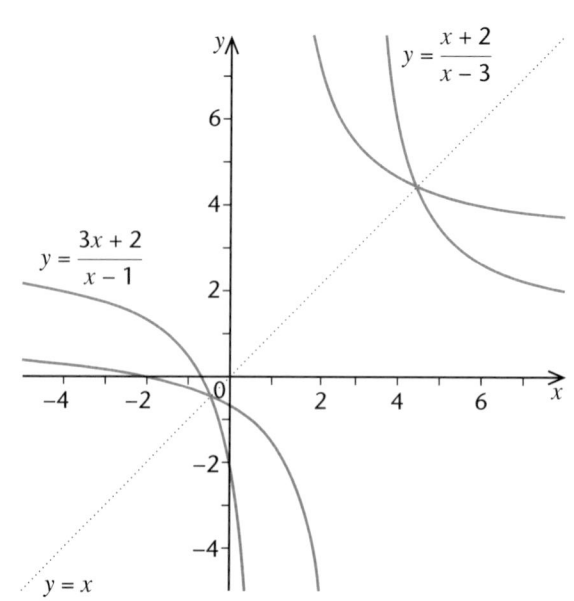

The modulus function

AQA C3
EDEXCEL C3
OCR C3
WJEC C3
CCEA C3

The notation $|x|$ is used to stand for the modulus of x. This is defined as

$$|x| = \begin{cases} x & \text{when} \quad x \geq 0 \quad \text{(when } x \text{ is positive, } |x| \text{ is just the same as } x\text{).}\\ -x & \text{when} \quad x < 0 \quad \text{(when } x \text{ is negative, } |x| \text{ is the same as } -x\text{).} \end{cases}$$

The expression $|x-a|$ can be interpreted as the distance between the numbers x and a on the number line. In this way, the statement $|x-a| < b$ means that the distance between x and a is less than b.

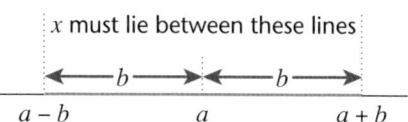

It follows that $a - b < x < a + b$.

The graph of $y = |x|$

It follows that the graph of $y = |x|$ is the same as the graph of $y = x$ for positive values of x. But, when x is negative, the corresponding part of the graph of $y = x$ must be reflected in the x-axis to give the graph of $y = |x|$.

$|x|$ is never negative so the graph of $y = |x|$ doesn't go below the x-axis anywhere.

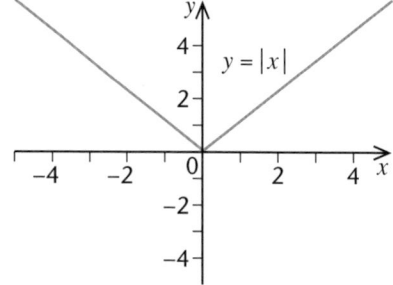

The graph of $y = |f(x)|$

The graph of $y = |f(x)|$ is the same as the graph of $y = f(x)$ for positive values of $f(x)$. But, when $f(x)$ is negative, the corresponding part of the graph of $y = f(x)$ must be reflected in the x-axis to give the graph of $y = |f(x)|$.

The graph of $y = 1/x$ goes below the x-axis for negative values of x.

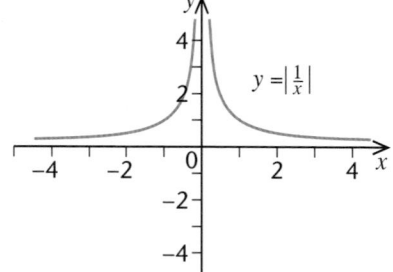

The diagram shows the graph of $y = \left| \dfrac{1}{x} \right|$.

Transforming graphs

AQA C3
EDEXCEL C3
OCR C3
WJEC C3
CCEA C3

The graph of some new function can often be obtained from the graph of a known function by applying a transformation. A summary of the standard transformations is given in the table.

You may need to apply a combination of transformations in some cases.

Known function	New function	Transformation
$y = f(x)$	$y = f(x) + a$	Translation through a units parallel to y-axis.
	$y = f(x - a)$	Translation through a units parallel to x-axis.
	$y = af(x)$	One-way stretch with scale factor a parallel to the y-axis.
	$y = f(ax)$	One-way stretch with scale factor $\frac{1}{a}$ parallel to the x-axis.

Examples

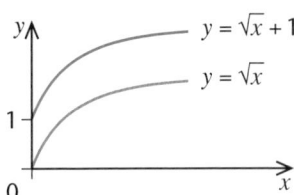 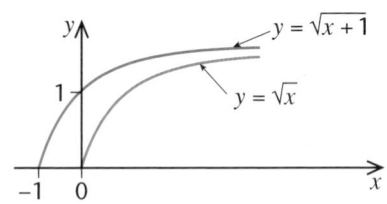

y = af(x) does not move points on the x-axis.

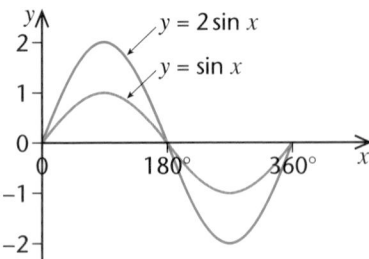 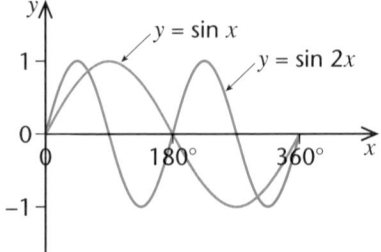

Sometimes you may have to perform more than one transformation.

Example

Describe how to transform the circle $x^2 + y^2 = 25$, with centre (0, 0) and radius 5, to the circle with centre (3, −8) and radius 5. Write down the new equation.

Key points from AS

Revise AS Circles 31

Translate 3 units parallel to the x axis (to the right) and −8 units parallel to the y axis (down).

$$x^2 + y^2 = 25 \rightarrow (x-3)^2 + (y+8)^2 = 25$$

The exponential function e^x

AQA	C3
EDEXCEL	C3
OCR	C3
WJEC	C3
CCEA	C3

Exponential functions are often used to represent, or model, patterns of:
• growth when $a > 1$
• decay when $a < 1$.

Key points from AS

*Revise AS
Exponential functions 43*

All scientific calculators will display values of e^x and 10^x. To display the value of e, for example, find e^1.

An exponential function is one where the variable is a power or exponent. For example, any function of the form $f(x) = a^x$, where a is a constant, is an exponential function.

The diagram shows some graphs of exponential functions for different values of a.

All of the graphs pass through the point (0,1) and each one has a different gradient at this point. The value of the gradient depends on a.

There is a particular value of a for which the curve has gradient 1 at (0,1). The letter e is used to stand for this value and the function $f(x) = e^x$ is often called the exponential function because of its special importance to the subject.

e = 2.718281828... It is an irrational number and so it cannot be written exactly.

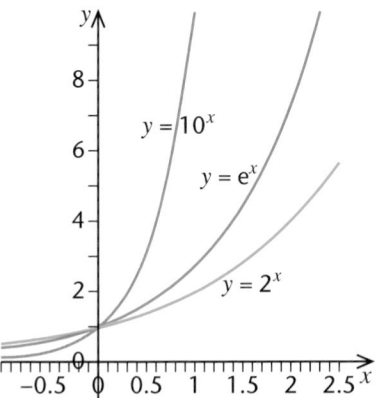

You may be asked to sketch other curves relating to e^x, for example

Notice $e^x > 0$ for all values of x.

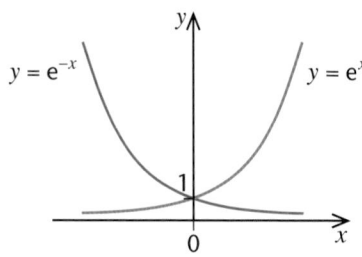

Reflection in y-axis.

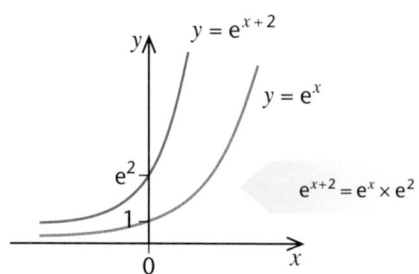

Translation 2 units left
or stretch, factor e^2, parallel to y-axis.

The natural logarithmic function, ln x

AQA	C3
EDEXCEL	C3
OCR	C3
WJEC	C3
CCEA	C3

Key points from AS

Revise AS Logarithms 43

If $y = e^x$ then $x = \log_e y$. This is usually written as $x = \ln y$.

Logarithms to the base e are called *natural logarithms*.

Your calculator has a key for natural logs, probably labelled ln.

$e^0 = 1 \Rightarrow \ln 1 = 0$
$e^1 = e \Rightarrow \ln e = 1$

In fact, $\ln e^{f(x)} = f(x)$, for example $\ln e^{x^2} = x^2$, $\ln e^{\sin x} = \sin x$.

Since the exponential and natural logarithm functions are inverses of each other, their graphs are symmetrical about the line $y = x$.

This is true for logs of any base.

The diagram shows that the domain of the natural logarithm function is given by $x > 0$.

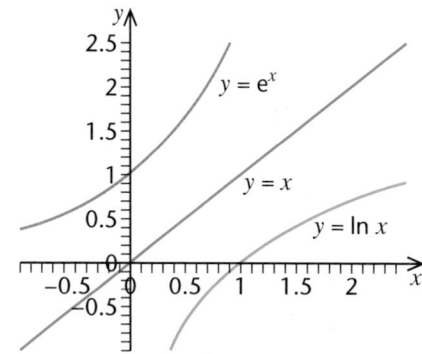

You may be asked to sketch other curves relating to $y = \ln x$, for example

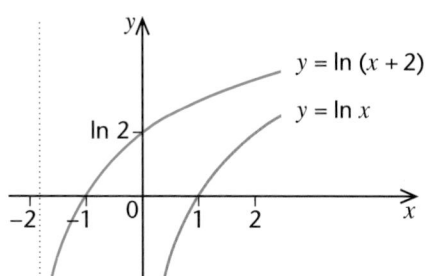

Translation 2 units to the left.

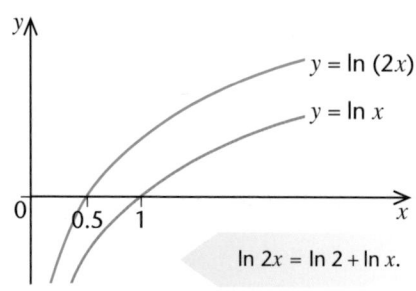

Stretch parallel to the x-axis factor $\frac{1}{2}$ or translation ln 2 units up.

Progress check

1 $f(x) = x^2 + 5$ and $g(x) = 2x + 3$.

(a) Write $fg(x)$ in terms of x.　　　　　(b) Find $fg(10)$.

(c) Find the values of x for which $fg(x) = gf(x)$.

2 Find the inverse of the function $f(x) = \dfrac{x+5}{x-2}$, $x \in \mathbb{R}$, $x \neq 2$ and state the domain of the inverse function.

3 (a) Sketch $y = (x-2)(x+2)$.

(b) Sketch $y = |x^2 - 4|$.

4 Solve $|x - 2| < 5$.

5 The diagram shows $y = f(x)$
On separate diagrams, sketch
(a) $y = f(x) + 1$
(b) $y = f(x + 1)$
(c) $y = f(2x)$
(d) $y = 2f(x)$
(e) $y = f(\tfrac{1}{2}x)$

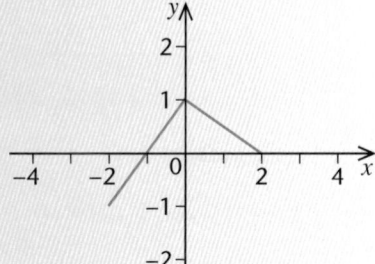

6 On the same diagram, sketch $y = x^2$, $y = (x-2)^2$, $y = -x^2$.

7 On the same diagram, sketch $y = e^x$, $y = e^{2x}$, $y = -e^x$.

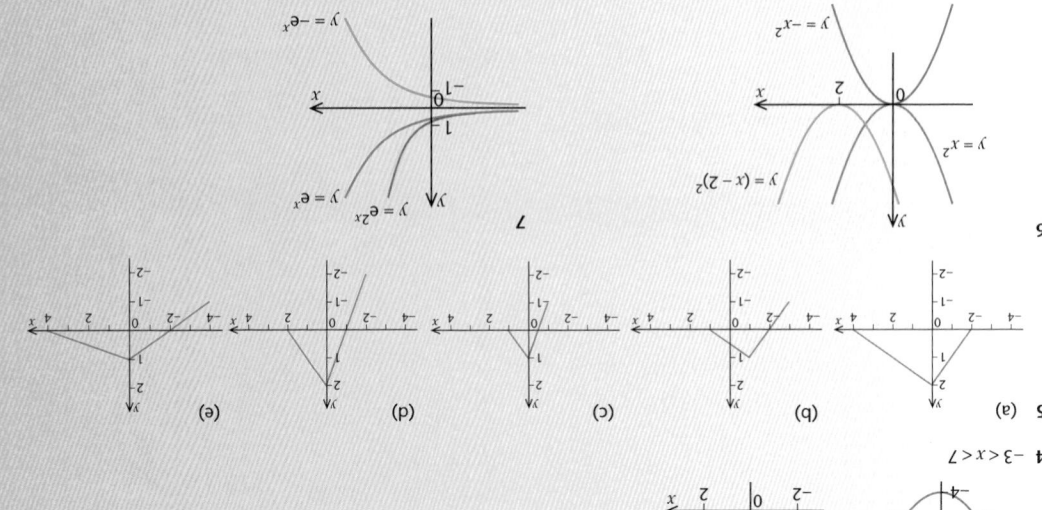

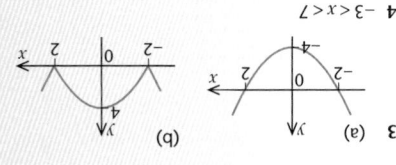

4 $-3 < x < 7$

2 $f(x) = \dfrac{2x+5}{x-1}$, $x \in \mathbb{R}$, $x \neq 1$.

1 (a) $fg(x) = 4x^2 + 12x + 14$ (b) $fg(10) = 534$ (c) $x = -0.08452, -5.915$.

1.3 Coordinate geometry

After studying this section you should be able to:

- *sketch curves in Cartesian and parametric form*
- *convert the equation of a curve between Cartesian and parametric form*

LEARNING SUMMARY

Cartesian and parametric form

AQA	C4
EDEXCEL	C4
OCR	C4
WJEC	C4
CCEA	C3

Key points from AS

- **Curve sketching**
  Revise AS page 23

If all the powers of x are even, the graph is symmetrical in the y-axis.

Look for values that make the denominator zero.

Cartesian form

A curve in Cartesian form is defined in terms of x and y only, **for example**
  $y = x^2 + 3x$ in which y is given **explicitly** in terms of x.
  $x^2 + 2xy + y^2 = 9$ in which the equation is given **implicitly**.

To sketch a function given in Cartesian form:

1 Find where the curve **crosses the axes** by putting $x = 0$ and $y = 0$.
2 Check for **symmetry**.
3 Check for **stationary points**.　　　Find when $\dfrac{dy}{dx} = 0$ and then investigate.
4 Check what happens for **large values** of x and y.
5 Check for **discontinuities**.

The parameter θ is often used when trigonometric functions are involved.

The curve is plotted by finding the values of x and y for various values of the parameter.

Parametric form

A curve is defined in **parametric form** by expressing x and y in terms of a third variable, for example $x = f(t)$, $y = g(t)$. In this case, t is the parameter.

Example
Plot the curve $x = 2t^2$, $y = 4t$ for $-3 \leqslant t \leqslant 3$ and find the Cartesian equation of the curve.

The curve is a parabola.

t	-3	-2	-1	0	1	2	3
x	18	8	2	0	2	8	18
y	-12	-8	-4	0	4	8	12

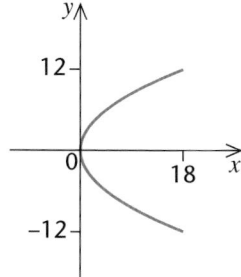

Eliminate t to obtain the Cartesian equation.

$$y = 4t \implies t = \frac{y}{4}.$$

Substituting into $x = 2t^2$ gives $x = 2\left(\dfrac{y}{4}\right)^2 = \dfrac{y^2}{8}$ so $y^2 = 8x$.

Key points from AS

- **Trigonometric identities**
  Revise AS page 53

When the parameter is θ, it may be necessary to use a trigonometric identity to find the Cartesian equation.

Key points from AS

- **Circle**
  Revise AS page 31

For example, when $x = a \cos \theta$ and $y = a \sin \theta$, use $\cos^2 \theta + \sin^2 \theta = 1$. This gives $\dfrac{x^2}{a^2} + \dfrac{y^2}{a^2} = 1$ or $x^2 + y^2 = a^2$.

This is a circle, centre (0, 0), radius a.

Also, when $x = a \cos \theta$, $y = b \sin \theta$, $\dfrac{x^2}{a^2} + \dfrac{y^2}{b^2} = 1$.

This curve is an ellipse.

Progress check

1. Sketch the curve $y = x^3 + 3x^2$.

2. A curve is defined parametrically by $x = 2t$, $y = \dfrac{2}{t}$.

   (a) Give the Cartesian equation of the curve.
   (b) Sketch the curve.

3. Describe the curve defined parametrically by $x = 5 \cos \theta$, $y = 5 \sin \theta$ and give its Cartesian equation.

3. Circle, centre (0, 0), radius 5; $x^2 + y^2 = 25$.

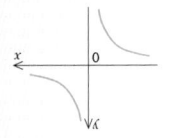

(b)

2. (a) $xy = 4$ (hyperbola)

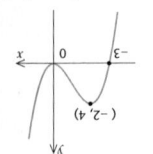

1

1.4 Trigonometry

After studying this section you should be able to:

- *understand the functions secant, cosecant and cotangent and be able to draw their graphs*
- *understand and use the Pythagorean identities*
- *understand and use the compound angle formulae for sine, cosine and tangent*
- *use identities to solve equations*

LEARNING SUMMARY

Secant, cosecant and cotangent

AQA   C3
EDEXCEL   C3
OCR   C3
WJEC   C3
NICCEA   C3

These trigonometric functions, commonly known as **sec**, **cosec** and **cot**, are defined from the more familiar sin, cos and tan functions as follows:

$$\sec x = \frac{1}{\cos x} \qquad \operatorname{cosec} x = \frac{1}{\sin x} \qquad \cot x = \frac{1}{\tan x}$$

Key points from AS
- **Trigonometry** page 50–53

Each function is periodic and has either line or rotational symmetry. The domain of each one must be restricted to avoid division by zero.
You need to be able to work in degrees or radians.
Remember $180° = \pi$ radians.

$y = \sec x$

- The period of sec is $360°$ (2π radians) to match the period of cos.
- Notice that $\sec x$ is undefined whenever $\cos x = 0$.
- The graph is symmetrical about every vertical line passing through a vertex.
- It has rotational symmetry of order 2 about the points in the x-axis corresponding to $90° \pm 180°n$ ($\frac{\pi}{2} \pm \pi n$).

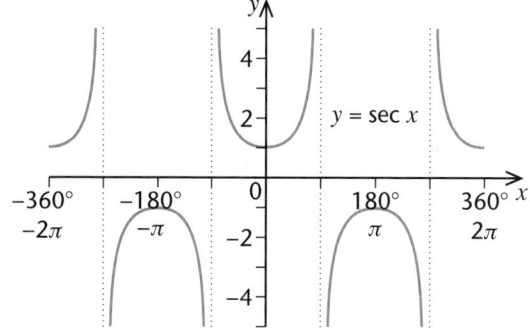

$y = \operatorname{cosec} x$

- The period of cosec is $360°$ (2π radians) to match the period of sin.
- Notice that $\operatorname{cosec} x$ is undefined whenever $\sin x = 0$.
- The graph is symmetrical about every vertical line passing through a vertex.
- It has rotational symmetry of order 2 about the points on the x-axis corresponding to $0°$, $\pm 180°$, $\pm 360°$, ..., $(0, \pm \pi, \pm 2\pi, ...)$.

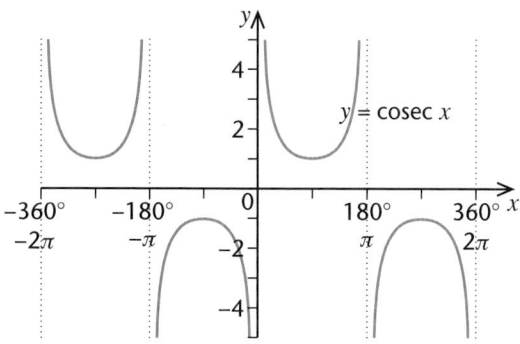

$y = \cot x$

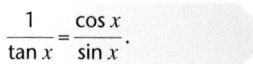

$$\frac{1}{\tan x} = \frac{\cos x}{\sin x}.$$

- The period of cot is $180°$ to match the period of tan.
- Notice that $\cot x$ is undefined whenever $\sin x = 0$.
- The graph has rotational symmetry of order 2 about the points on the x-axis corresponding to $0°$, $\pm 90°$, $\pm 180°$, ..., $(0, \pm \frac{\pi}{2}, \pm \pi, ...)$.

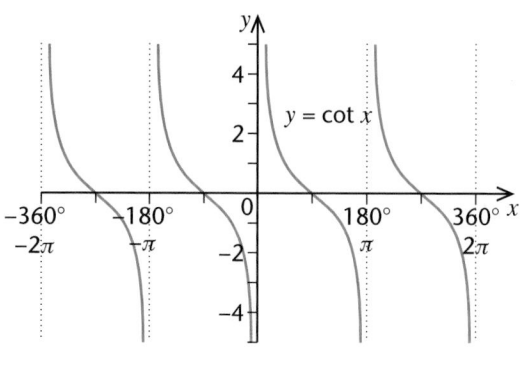

Inverse trigonometric functions

AQA — C3
EDEXCEL — C3
OCR — C3
WJEC — C3
CCEA — C3

See page 23 to find out about inverse functions.

You need to know these graphs and to understand the need for restricting the domains.

The sine, cosine and tangent functions are all many–one and so do not have inverses on their full domains. However, it is possible to restrict their domains so that each one has an inverse. The graphs of these inverse functions are given below.

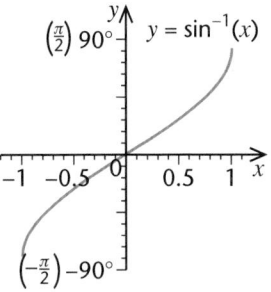 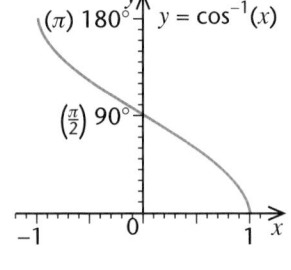 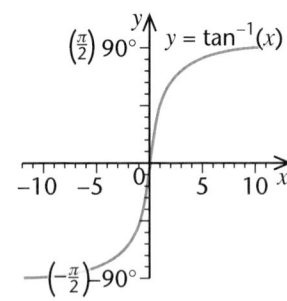

$f(x) = \sin x, \; -90° \leqslant x \leqslant 90°$
$\left(-\frac{\pi}{2} \leqslant x \leqslant \frac{\pi}{2}\right)$
$\Rightarrow f^{-1}(x) = \sin^{-1}x, \; -1 \leqslant x \leqslant 1.$

$f(x) = \cos x, \; 0° \leqslant x \leqslant 180°$
$(0 \leqslant x \leqslant \pi)$
$\Rightarrow f^{-1}(x) = \cos^{-1}x, \; -1 \leqslant x \leqslant 1.$

$f(x) = \tan x, \; -90° < x < 90°$
$\left(-\frac{\pi}{2} < x < \frac{\pi}{2}\right)$
$\Rightarrow f^{-1}(x) = \tan^{-1}x, \; x \in \mathbb{R}.$

This is the notation used for inverse trig' functions on your calculator.

But some texts use:
arcsin for $\sin^{-1}$
arccos for $\cos^{-1}$
arctan for $\tan^{-1}$.

$\sin^{-1}x$ means:
'the angle whose sine is x'
e.g. $\sin^{-1}0.5 = 30°$.

In radians $\sin^{-1}0.5 = \frac{\pi}{6}$.

$\cos^{-1}x$ means:
'the angle whose cosine is x'
e.g. $\cos^{-1}0.5 = 60°$.
In radians $\cos^{-1}0.5 = \frac{\pi}{3}$.

$\tan^{-1}x$ means:
'the angle whose tangent is x'
e.g. $\tan^{-1}1 = 45°$.
In radians $\tan^{-1}1 = \frac{\pi}{4}$.

When you use the functions $\sin^{-1}$, $\cos^{-1}$ and $\tan^{-1}$ on your calculator, the value given is called the **principal value** (PV).

Trigonometric identities

AQA — C3
EDEXCEL — C3
OCR — C3
WJEC — C3
CCEA — C3

The second result can be obtained from the first by dividing throughout by $\cos^2\theta$.
The third result can be obtained in a similar way, by dividing throughout by $\sin^2\theta$.

You should *learn* the identities listed below. It is so much easier to simplify expressions, establish new identities and solve equations if you have a firm grasp of these basic results.

- The **Pythagorean identities** are:

$$\cos^2\theta + \sin^2\theta \equiv 1$$

$$1 + \tan^2\theta \equiv \sec^2\theta$$

$$\cot^2\theta + 1 \equiv \operatorname{cosec}^2\theta$$

AQA — C4
EDEXCEL — C3
OCR — C3
WJEC — C4
CCEA — C4

- The **compound angle identities** for $(A + B)$ are:

$$\sin(A + B) \equiv \sin A \cos B + \cos A \sin B$$

$$\cos(A + B) \equiv \cos A \cos B - \sin A \sin B$$

$$\tan(A + B) \equiv \frac{\tan A + \tan B}{1 - \tan A \tan B}.$$

These results can be worked out from the ones for $(A + B)$. Just replace every '+' with a '−' and vice versa. This makes it easier to learn the results.

The corresponding results for $(A - B)$ are:

$$\sin(A - B) \equiv \sin A \cos B - \cos A \sin B$$

$$\cos(A - B) \equiv \cos A \cos B + \sin A \sin B$$

$$\tan(A - B) \equiv \frac{\tan A - \tan B}{1 + \tan A \tan B}.$$

You can obtain these results from the compound angle formulae by setting $A = B$.

- The **double angle identities** are:

$$\sin 2A \equiv 2 \sin A \cos A$$

$$\cos 2A \equiv \cos^2 A - \sin^2 A$$

$$\equiv 2 \cos^2 A - 1$$

$$\equiv 1 - 2 \sin^2 A$$

These variations are very useful – it is worth knowing them.

$$\tan 2A \equiv \frac{2 \tan A}{1 - \tan^2 A}.$$

- The **half angle identities** are:

These results will be useful for integration later in the course.

$$\sin^2 \tfrac{1}{2} A \equiv \tfrac{1}{2} (1 - \cos A)$$

$$\cos^2 \tfrac{1}{2} A \equiv \tfrac{1}{2} (1 + \cos A).$$

Proving identities

AQA	C3/C4
EDEXCEL	C3
OCR	C3
WJEC	C3/C4
CCEA	C3/C4

The basic technique for proving a given identity is:

- Start with the expression on one side of the identity.
- Use known identities to replace some part of it with an equivalent form.
- Simplify the result and compare with the expression on the other side.

This is a skill. The more you do, the better you will become.

With practice you get to know which parts to replace so that the required result is established.

Example

Prove that $\dfrac{\cos 2A}{\cos A - \sin A} \equiv \cos A + \sin A$.

Starting with the LHS:

$$\frac{\cos 2A}{\cos A - \sin A} \equiv \frac{\cos^2 A - \sin^2 A}{\cos A - \sin A}$$

Using one of the double angle results.

$$\equiv \frac{(\cos A + \sin A)(\cos A - \sin A)}{\cos A - \sin A}$$

Using the difference of two squares.

$$\equiv \cos A + \sin A \equiv \text{RHS}.$$

Cancelling the common factor.

It is often a good idea to replace sec, cosec and cot with sin, cos and tan at the start.

Example

Prove that $\sec^2 A \equiv \dfrac{\operatorname{cosec} A}{\operatorname{cosec} A - \sin A}$.

It's often easier to start with the more complicated looking side and try to simplify it.

$$\text{RHS} \equiv \frac{\dfrac{1}{\sin A}}{\dfrac{1}{\sin A} - \sin A} \equiv \frac{1}{1 - \sin^2 A}$$

Replacing cosec A with $\dfrac{1}{\sin A}$ and multiplying the numerator and denominator by sin A.

$$\equiv \frac{1}{\cos^2 A} \equiv \sec^2 A \equiv \text{LHS}.$$

Using one of the Pythagorean identities and the definition of sec A.

Trigonometric equations

AQA	C3/C4
EDEXCEL	C3
OCR	C3
WJEC	C4
CCEA	C3/C4

Identities may be used to re-write an equation in a form that is easier to solve.

Example Solve the equation $\cos 2x + \sin x + 1 = 0$ for $-\pi \leqslant x \leqslant \pi$.

Replacing $\cos 2x$ with $1 - 2\sin^2 x$ gives $1 - 2\sin^2 x + \sin x + 1 = 0$

$$\Rightarrow 2\sin^2 x - \sin x - 2 = 0.$$

> You need to recognise that this is a quadratic in $\sin x$.

Using the quadratic formula gives $\sin x = \dfrac{1 \pm \sqrt{1+16}}{4} = \dfrac{1 \pm \sqrt{17}}{4}$

$$\Rightarrow \sin x = 1.28 \ldots \text{ (no solutions) or } \sin x = -0.78077 \ldots .$$

> $-1 \leqslant \sin x \leqslant 1$ for all values of x.

Using $\sin^{-1}$ gives $\qquad x = -0.8959$ radians to 4 d.p.

> Set calculator to radian mode.

Another solution in the interval $-\pi \leqslant x \leqslant \pi$ is $-\pi + 0.8959$ giving $x = -2.2457$ radians to 4 d.p.

The diagram shows there are no other solutions in this interval.

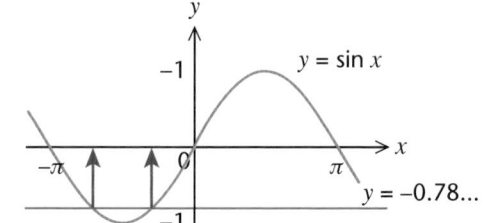

AQA	C4
EDEXCEL	C3
OCR	C3
WJEC	C4
CCEA	C4

> **KEY POINT**
>
> An important result is that the expression $a\cos\theta + b\sin\theta$ can be written in any of the equivalent forms $r\cos(\theta \pm \alpha)$ or $r\sin(\theta \pm \alpha)$ for a suitable choice of r and α. The convention is to take $r > 0$.

One application of the result is to solve equations of the form $a\cos\theta + b\sin\theta = c$.

Example Express $3\cos\theta - 4\sin\theta$ in the form $r\cos(\theta + \alpha)$, $r > 0$. Use the result to solve the equation $3\cos\theta - 4\sin\theta = 2$ for $-\pi \leqslant \theta \leqslant \pi$.

> You need to expand $r\cos(\theta + \alpha)$ and match corresponding parts of the identity.

$$r\cos(\theta + \alpha) \equiv 3\cos\theta - 4\sin\theta$$

$$\Rightarrow r\cos\theta\cos\alpha - r\sin\theta\sin\alpha \equiv 3\cos\theta - 4\sin\theta$$

$$\Rightarrow r\cos\alpha = 3 \text{ and } r\sin\alpha = 4$$

> $\dfrac{r\sin\alpha}{r\cos\alpha} = \dfrac{4}{3}$
>
> $\Rightarrow \tan\alpha = \dfrac{4}{3}$

$$\Rightarrow r = \sqrt{3^2 + 4^2} = 5 \text{ and } \alpha = \tan^{-1}\left(\tfrac{4}{3}\right) = 0.9273^c \text{ to 4 d.p.}$$

$$\Rightarrow 3\cos\theta - 4\sin\theta \equiv 5\cos(\theta + 0.9273^c).$$

Using this result in the equation $3\cos\theta - 4\sin\theta = 2$ gives

$$5\cos(\theta + 0.9273^c) = 2$$

$$\Rightarrow \cos(\theta + 0.9273^c) = 0.4.$$

This gives $\theta + 0.9273^c = \pm 1.1593^c \Rightarrow \theta = 0.232^c$ or $\theta = -2.087^c$ to 3 d.p.

Progress check

1 Prove the identity $\sin 2\theta \equiv \dfrac{2\tan\theta}{1 + \tan^2\theta}$.

2 Solve the equation $\cos 2x - \sin x = 0$ for $0° \leqslant x \leqslant 360°$.

3 (a) Write $5\sin x + 12\cos x$ in the form $r\sin(x + \alpha)$ where $0 < \alpha < \dfrac{\pi}{2}$.

    (b) Solve the equation $5\sin x + 12\cos x = 7$ for $-\pi \leqslant x \leqslant \pi$.

3 (a) $13\sin(x + 1.176^c)$ (b) $x = -0.607^c$ or $x = 1.397^c$ to 3 d.p.

2 $x = 30°, 150°$ or $270°$

1 RHS $= \dfrac{2\tan\theta}{\sec^2\theta} = 2\tan\theta \times \cos^2\theta = 2\sin\theta\cos\theta = \sin 2\theta \equiv$ LHS.

1.5 Differentiation

After studying this section you should be able to:

- *use the chain rule to differentiate composite functions*
- *differentiate functions based on e^x and $\ln x$*
- *differentiate trigonometric functions*
- *differentiate using the product and quotient rules*
- *understand and use the relation $\dfrac{dy}{dx} = 1 \div \dfrac{dx}{dy}$*
- *find derivatives of functions defined parametrically and implicitly*

The chain rule

AQA C3
EDEXCEL C3
OCR C3
WJEC C3
NICCEA C3

The **chain rule** can be written as $\dfrac{dy}{dx} = \dfrac{dy}{du} \times \dfrac{du}{dx}$, where u is a function of x.

It shows how to differentiate a composite function by separating the process into simpler stages and combining the results.

Example Differentiate (a) $y = (3x + 2)^{10}$

(b) $y = \sqrt{x^2 - 9}$.

Key points from AS

- **Differentiation 34, 35**
  Revise AS pages 34, 55

(a) Taking $u = 3x + 2$ gives $y = u^{10}$ and $\dfrac{dy}{du} = 10u^9$, so $\dfrac{dy}{du} = 10(3x + 2)^9$.

It also gives $\dfrac{du}{dx} = 3$.

So, from the chain rule: $\dfrac{dy}{dx} = 10(3x + 2)^9 \times 3 = 30(3x + 2)^9$.

(b) Taking $u = x^2 - 9$ gives $y = \sqrt{u} = u^{\frac{1}{2}}$, and $\dfrac{dy}{du} = \frac{1}{2} u^{-\frac{1}{2}}$, so $\dfrac{dy}{du} = \frac{1}{2}(x^2 - 9)^{-\frac{1}{2}}$.

It also gives $\dfrac{du}{dx} = 2x$.

So, from the chain rule: $\dfrac{dy}{dx} = \frac{1}{2}(x^2 - 9)^{-\frac{1}{2}} \times 2x = \dfrac{x}{\sqrt{x^2 - 9}}$.

The chain rule can also be used to establish results for **connected rates of change**.

OCR C3

For example, the rate of change of the volume of a sphere can be written as $\dfrac{dv}{dt}$.

The corresponding rate of change of the radius of the sphere can be written as $\dfrac{dr}{dt}$. Using the chain rule, the connection between these rates of change is given by:

$$\frac{dv}{dt} = \frac{dv}{dr} \times \frac{dr}{dt}.$$

Also, since $v = \frac{4}{3}\pi r^3$, it follows that $\dfrac{dv}{dr} = 4\pi r^2$, so, the connection can now be written as $\dfrac{dv}{dt} = 4\pi r^2 \dfrac{dr}{dt}$.

The exponential function e^x

AQA   C3
EDEXCEL   C3
OCR   C3
WJEC   C3
CCEA   C3

The exponential function was defined on page 26. It has the special property that it is unchanged by differentiation. So, if $y = e^x$ then $\dfrac{dy}{dx} = e^x$.

Constant multiples of the exponential function behave in the same way, so if

$y = ae^x$ then $\dfrac{dy}{dx} = ae^x$ where a is any constant.

Examples
Differentiate with respect to x:

(a) $y = 2e^x$ (b) $y = x^3 + e^x$ (c) $f(x) = \dfrac{1}{x} - 3e^x$

(a) $\dfrac{dy}{dx} = 2e^x$ (b) $\dfrac{dy}{dx} = 3x^2 + e^x$ (c) $f'(x) = -\dfrac{1}{x^2} - 3e^x$

You may need to use the chain rule.

Examples
Differentiate with respect to x:

(a) $y = e^{3x+2}$ (b) $y = e^{4x^2}$

(a) Let $u = 3x + 2$ so that $y = e^u$

$\dfrac{du}{dx} = 3$ and $\dfrac{dy}{du} = e^u$

Using the chain rule

$\dfrac{dy}{dx} = \dfrac{dy}{du} \times \dfrac{du}{dx} = e^u \times 3 = 3\,e^{3x+2}$

(b) Let $u = 4x^2$, so that $y = e^u$

$\dfrac{du}{dx} = 8x$ and $\dfrac{dy}{du} = e^u$

$\dfrac{dy}{dx} = \dfrac{dy}{du} \times \dfrac{du}{dx} = e^u \times 8x = 8xe^{4x^2}$

In general, it can be shown using the chain rule

> $f'(x)$ is the derivative, with respect to x, of $f(x)$.

KEY POINT

If $y = e^{f(x)}$ then $\dfrac{dy}{dx} = f'(x)e^{f(x)}$.

In particular, it is useful to remember that if $y = e^{ax+b}$, then $\dfrac{dy}{dx} = ae^{ax+b}$

Logarithmic functions $\ln x$ and $\ln f(x)$

AQA   C3
EDEXCEL   C3
OCR   C3
WJEC   C3
CCEA   C3

KEY POINT

If $y = \ln x$ then $\dfrac{dy}{dx} = \dfrac{1}{x}$.

This result is easily extended to functions of the form $y = \ln(kx)$ and $y = \ln(x^k)$ using the laws of logarithms.

If $y = \ln(kx)$, then $y = \ln k + \ln x$

giving $\dfrac{dy}{dx} = \dfrac{1}{x} + 0 = \dfrac{1}{x}$.

> Since $\ln k$ is a constant, its derivative is zero.

For example, if $y = \ln(5x)$, then $\dfrac{dy}{dx} = \dfrac{1}{x}$.

If $y = \ln (x^k)$, then $y = k \ln x$ giving $\dfrac{dy}{dx} = k \times \dfrac{1}{x} = \dfrac{k}{x}$.

For example, if $y = \ln\left(\dfrac{1}{x^2}\right) = \ln(x^{-2}) = -2 \ln x$, then $\dfrac{dy}{dx} = -2 \times \dfrac{1}{x} = -\dfrac{2}{x}$.

Always check whether a logarithmic function can be simplified before attempting to differentiate.

You may need to use the chain rule to differentiate $y = \ln f(x)$.

Example

> This function cannot be simplified.

Differentiate $y = \ln(x^2 + 3x)$.

Let $u = x^2 + 3x$ so that $y = \ln u$

$$\frac{du}{dx} = 2x + 3 \quad \text{and} \quad \frac{dy}{du} = \frac{1}{u} = \frac{1}{x^2 + 3x}$$

Using the chain rule

$$\frac{dy}{dx} = \frac{dy}{du} \times \frac{du}{dx} = \frac{1}{x^2 + 3x} \times (2x + 3)$$

> Notice that $2x + 3$ is the derivative of $x^2 + 3x$.

$$= \frac{2x + 3}{x^2 + 3x}$$

In general, it can be shown using the chain rule

> This can be used directly without showing the chain rule working.

$$\frac{d}{dx} \ln(f(x)) = \frac{f'(x)}{f(x)}.$$

KEY POINT

Progress check

Find $\dfrac{dy}{dx}$ in each of the following questions.

1 $y = (4 - 3x)^6$

2 $y = \sqrt{2x + 1}$

3 $y = e^{3x+1} + 3x$

4 $y = \ln(7x) + \ln(x^4)$

5 $y = \ln(x^2 - 3)$

5 $\dfrac{2x}{x^2 - 3}$

4 $\dfrac{5}{x}$

3 $3e^{3x+1} + 3$

2 $(2x + 1)^{-\frac{1}{2}}$

1 $-18(4 - 3x)^5$

Trigonometric functions

AQA   C3
EDEXCEL   C3
OCR   C4
WJEC   C3
CCEA   C3

You should learn the derivatives of the three main **trigonometric functions**:

$$\frac{d}{dx}(\sin x) = \cos x \qquad\qquad \frac{d}{dx}\sin(ax+b) = a\cos(ax+b)$$

$$\frac{d}{dx}(\cos x) = -\sin x \qquad\qquad \frac{d}{dx}\cos(ax+b) = -a\sin(ax+b)$$

$$\frac{d}{dx}(\tan x) = \sec^2 x \qquad\qquad \frac{d}{dx}\tan(ax+b) = a\sec^2(ax+b)$$

When differentiating, the angle must be in radians.

Example

Find the exact value of the gradient of $y = \cos 2x$ when $x = \dfrac{\pi}{6}$.

$$y = \cos 2x \implies \frac{dy}{dx} = -2\sin 2x$$

When $x = \dfrac{\pi}{6}$, $\quad \dfrac{dy}{dx} = -2\sin\dfrac{\pi}{3} = -2 \times \dfrac{\sqrt{3}}{2} = -\sqrt{3}$

The **chain rule** can often be used to differentiate expressions involving trigonometric functions.

Example

Differentiate: (a) $y = 5\sin^4 x$ (b) $y = e^{\tan 4x}$ (c) $y = \ln(\sin x)$

$\sin^4 x = (\sin x)^4$.

(a)   Let $u = \sin x$ so that $y = 5u^4$

     Then $\dfrac{du}{dx} = \cos x$ and $\dfrac{dy}{du} = 20u^3 = 20(\sin x)^3 = 20\sin^3 x$

With practice you should be able to do the working mentally.

     By the chain rule:

$$\frac{dy}{dx} = \frac{dy}{du} \times \frac{du}{dx} = 20\sin^3 x \times (\cos x)$$

$$= 20\cos x \sin^3 x.$$

Although not essential, it is usual to write an expression with the least complicated factors first.

(b)   Let $u = \tan 4x$ so that $y = e^u$

$$\frac{du}{dx} = 4\sec^2 4x \text{ and } \frac{dy}{du} = e^u = e^{\tan 4x}.$$

$$\therefore \quad \frac{dy}{dx} = e^{\tan 4x} \times 4\sec^2 4x$$

$$= 4\sec^2 4x\, e^{\tan 4x}.$$

In general, by the chain rule:
$$\frac{d}{dx}(e^{f(x)}) = f'(x)e^{f(x)}.$$

(c)   Use the result $y = \ln f(x) \implies \dfrac{dy}{dx} = \dfrac{f'(x)}{f(x)}$

$$y = \ln(\sin x)$$

$$\frac{dy}{dx} = \frac{\cos x}{\sin x}$$

$$= \cot x.$$

$f(x) = \sin x,\ f'(x) = \cos x.$

The product rule

AQA · C3
EDEXCEL · C3
OCR · C3
WJEC · C3
NICCEA · C3

Expressions written as the product of two functions, $f(x) \times g(x)$, can be differentiated using the **product rule**.

> **KEY POINT**
>
> If $y = uv$, where $u = f(x)$ and $v = g(x)$, then $\dfrac{dy}{dx} = u\dfrac{dv}{dx} + v\dfrac{du}{dx}$.

Example
Differentiate with respect to x: (a) $y = x^2 \sin 4x$ (b) $y = x^3 e^{-x}$

With practice, the side working can be done mentally.

(a) Let $y = uv$, where $u = x^2$ and $v = \sin 4x$ Side working

$$\frac{dy}{dx} = u\frac{dv}{dx} + v\frac{du}{dx} \qquad u = x^2 \Rightarrow \frac{du}{dx} = 2x$$

$$= x^2 \times 4\cos 4x + \sin 4x \times 2x \qquad v = \sin 4x \Rightarrow \frac{dv}{dx} = 4\cos 4x$$

$$= 2x(2x\cos 4x + \sin 4x)$$

Tidy the answer by taking out factors if possible.

(b) Let $y = uv$, where $u = x^3$ and $v = e^{-x}$ Side working

$$\frac{dy}{dx} = x^3 \times (-e^{-x}) + e^{-x} \times 3x^2 \qquad u = x^3 \Rightarrow \frac{du}{dx} = 3x^2$$

$$= x^2 e^{-x}(-x + 3) \qquad v = e^{-x} \Rightarrow \frac{dv}{dx} = -e^{-x}$$

The quotient rule

AQA · C3
EDEXCEL · C3
OCR · C3
WJEC · C3
NICCEA · C3

Expressions written in the form $\dfrac{f(x)}{g(x)}$ can be differentiated using the **quotient rule**.

> **KEY POINT**
>
> If $y = \dfrac{u}{v}$ where $u = f(x)$ and $v = g(x)$, then $\dfrac{dy}{dx} = \dfrac{v\dfrac{du}{dx} - u\dfrac{dv}{dx}}{v^2}$.

Before using the quotient rule, check whether the expression can be simplified first, such as in
$$y = \frac{4x^2 + 3x + 2}{x}.$$

Example
Find $\dfrac{dy}{dx}$ when $y = \dfrac{e^{2x}}{x^3}$.

Side working:

$$\frac{dy}{dx} = \frac{v\dfrac{du}{dx} - u\dfrac{dv}{dx}}{v^2}$$

$$= \frac{x^3(2e^{2x}) - e^{2x}(3x^2)}{(x^3)^2} \qquad u = e^{2x} \Rightarrow \frac{du}{dx} = 2e^{2x}$$

Take out factors.

$$= \frac{x^2 e^{2x}(2x - 3)}{x^6} \qquad v = x^3 \Rightarrow \frac{dv}{dx} = 3x^2$$

Simplify.

$$= \frac{e^{2x}(2x - 3)}{x^4}$$

Example

Find the coordinates of the stationary point on the curve $y = \dfrac{\ln x}{x}$.

Side working:

Key points from AS

- **Stationary points**
  Revise AS page 35

$$\frac{dy}{dx} = \frac{x \times \dfrac{1}{x} - \ln x \times 1}{x^2}$$

$$= \frac{1 - \ln x}{x^2}$$

$$u = \ln x \Rightarrow \frac{du}{dx} = \frac{1}{x}$$

$$v = x \Rightarrow \frac{dv}{dx} = 1$$

$\dfrac{dy}{dx} = 0$ when $\ln x = 1$, i.e. when $x = e$

$\ln e = 1$.

When $x = e$, $y = e^{-1}$, so the stationary point is at (e, e^{-1}).

Alternative formats of product and quotient rules when $u = f(x)$ and $v = g(x)$:

These formats are sometimes quoted. Learn and use the ones you find easiest to remember.

Product rule: If $y = f(x) \times g(x)$, then $\dfrac{dy}{dx} = f'(x)g(x) + f(x)g'(x)$

Quotient rule: If $y = \dfrac{f(x)}{g(x)}$, then $\dfrac{dy}{dx} = \dfrac{f'(x)g(x) - f(x)g'(x)}{[g(x)]^2}$

Using the result $\dfrac{dy}{dx} = 1 \div \dfrac{dx}{dy}$

AQA	C3
EDEXCEL	C3
OCR	C3
WJEC	C3
CCEA	C3

$$\frac{dy}{dx} = 1 \div \frac{dx}{dy} = \frac{1}{\dfrac{dx}{dy}}.$$

KEY POINT

For example, if $x = 4y + 2$, then $\dfrac{dx}{dy} = 4$ and $\dfrac{dy}{dx} = \dfrac{1}{\dfrac{dx}{dy}} = \dfrac{1}{4}$.

EDEXCEL	C3
WJEC	C3

The relationship is very useful for differentiating inverse trigonometrical functions.

Example

This could be written:
Find $\dfrac{dy}{dx}$ when $x = \sin y$.

Find $\dfrac{dy}{dx}$ when   (a) $y = \sin^{-1} x$ (b) $y = \tan^{-1} x$

(a) $y = \sin^{-1} x \Rightarrow x = \sin y$

Use $\cos^2 y + \sin^2 y = 1$.

$$\frac{dx}{dy} = \cos y = \sqrt{1 - \sin^2 y} = \sqrt{1 - x^2}$$

$$\frac{dy}{dx} = \frac{1}{\dfrac{dx}{dy}} = \frac{1}{\sqrt{1 - x^2}}$$

> Use $1 + \tan^2 y = \sec^2 y$.

(b) $y = \tan^{-1} x \Rightarrow x = \tan y$

$$\frac{dx}{dy} = \sec^2 y = 1 + \tan^2 y = 1 + x^2$$

$$\frac{dy}{dx} = \frac{1}{\dfrac{dx}{dy}} = \frac{1}{1 + x^2}$$

> EDEXCEL   C4

It is also useful for differentiating $y = a^x$ as follows.

Let $y = a^x$ and take logs to the base e of both sides.

> $\ln a^x = x \ln a$ by third law of logarithms.

$$y = a^x \Rightarrow \ln y = \ln a^x = x \ln a$$

$$x = \frac{1}{\ln a} \ln y$$

$$\frac{dx}{dy} = \frac{1}{\ln a} \frac{1}{y} \Rightarrow \frac{dy}{dx} = \frac{1}{\dfrac{dx}{dy}} = y \ln a = a^x \ln a$$

> **KEY POINT**
>
> If $y = a^x$, then $\dfrac{dy}{dx} = a^x \ln a$.

Parametric functions

First derivative

> AQA   C4
> EDEXCEL   C4
> OCR   C4
> WJEC   C3
> CCEA   C4

> **KEY POINT**
>
> If x and y are each expressed in terms of a **parameter** t,
> then $\dfrac{dy}{dx} = \dfrac{dy}{dt} \times \dfrac{dt}{dx}$.   Remember that $\dfrac{dt}{dx} = \dfrac{1}{dx/dt}$.

Example

A curve is defined parametrically by $x = 3t^2$, $y = 4t$.

(a) Find $\dfrac{dy}{dx}$ when $y = 8$ (b) Find the value of $\dfrac{d^2y}{dx^2}$ when $t = -1$.

(a) $x = 3t^2 \Rightarrow \dfrac{dx}{dt} = 6t \Rightarrow \dfrac{dt}{dx} = \dfrac{1}{6t}$

$$y = 4t \Rightarrow \frac{dy}{dt} = 4$$

$$\therefore \quad \frac{dy}{dx} = \frac{dy}{dt} \times \frac{dt}{dx} = 4 \times \frac{1}{6t} = \frac{2}{3t}$$

> Find the value of t when $y = 8$.

When $y = 8$, $4t = 8$, i.e. $t = 2$; when $t = 2$, $\dfrac{dy}{dx} = \dfrac{2}{3t} = \dfrac{2}{6} = \dfrac{1}{3}$.

Second derivative

> WJEC   C3
> CCEA   C4

(b) $\dfrac{d^2y}{dx^2} = \dfrac{d}{dx}\left(\dfrac{2}{3t}\right)$

$$= \frac{d}{dt}\left(\frac{2}{3t}\right) \times \frac{dt}{dx}$$

> If $\dfrac{dy}{dx} = h(t)$, then
>
> $\dfrac{d^2y}{dx^2} = \dfrac{d}{dt}(h(t)) \times \dfrac{dt}{dx}$

$$= -\frac{2}{3} t^{-2} \times \frac{1}{6t} = -\frac{1}{9t^3}$$

When $t = -1$, $\dfrac{d^2y}{dx^2} = \dfrac{1}{9}$

Implicit functions

First derivative

AQA	C4
EDEXCEL	C4
OCR	C4
WJEC	C3
CCEA	C4

> To find $\dfrac{dy}{dx}$ when an equation in x and y is given **implicitly**, differentiate each term with respect to x, remembering that $\dfrac{d}{dx}(f(y)) = \dfrac{d}{dy}f(y) \times \dfrac{dy}{dx}$.

Example

The point $P(2, 6)$ lies on the circle $x^2 + y^2 + 2x - 4y - 20 = 0$.

(a) Show that $\dfrac{dy}{dx} = \dfrac{x+1}{2-y}$

(b) Calculate the gradient of the tangent at P.

(c) Calculate the value of $\dfrac{d^2y}{dx^2}$ at P.

> Differentiate term by term with respect to x
> $\dfrac{d}{dx}(y) = \dfrac{dy}{dx}$.

(a) $2x + 2y\dfrac{dy}{dx} + 2 - 4\dfrac{dy}{dx} + 0 = 0$

> Divide throughout by 2 to simplify the equation.

$$x + y\dfrac{dy}{dx} + 1 - 2\dfrac{dy}{dx} = 0 \qquad (1)$$

> Get the terms in $\dfrac{dy}{dx}$ on one side of the equation and all the other terms on the other side.

$$x + 1 = 2\dfrac{dy}{dx} - y\dfrac{dy}{dx}$$

> Take out $\dfrac{dy}{dx}$ as a factor.

$$\Rightarrow x + 1 = \dfrac{dy}{dx}(2 - y)$$

$$\therefore \quad \dfrac{dy}{dx} = \dfrac{x+1}{2-y}$$

> The value of $\dfrac{dy}{dx}$ when $x = 2$ gives the gradient of the tangent at $(2, 6)$

(b) At P, $x = 2$ and $y = 6$,

so $\dfrac{dy}{dx} = \dfrac{3}{-4} = -\dfrac{3}{4}$.

The gradient of the tangent at P is $-\dfrac{3}{4}$.

Second derivative

| CCEA | C4 |

> To find $\dfrac{d^2y}{dx^2}$, differentiate (1) with respect to x.

(c) $\qquad x + y\dfrac{dy}{dx} + 1 - 2\dfrac{dy}{dx} = 0 \qquad$ from (1) above

> Use the product rule to differentiate $y\dfrac{dy}{dx}$.

$$1 + \left(y\dfrac{d^2y}{dx^2} + \dfrac{dy}{dx} \times \dfrac{dy}{dx}\right) - 2\dfrac{d^2y}{dx^2} = 0$$

$$1 + (y-2)\dfrac{d^2y}{dx^2} + \left(\dfrac{dy}{dx}\right)^2 = 0 \qquad (2)$$

At P, $y = 6$ and $\dfrac{dy}{dx} = -\dfrac{3}{4}$ $\qquad$ from part (b)

Substituting in (2)

$$1 + 4\dfrac{d^2y}{dx^2} + \left(-\dfrac{3}{4}\right)^2 = 0$$

$$\Rightarrow \dfrac{d^2y}{dx^2} = -\dfrac{25}{64}$$

Progress check

1 Differentiate with respect to x:

(a) $y = 2 \sin 6x$ (b) $y = 4 \cos^3 x$ (c) $y = 2 \tan 3x$

(d) $y = e^{\sin x}$ (e) $y = \sin(x^3)$

Use a trigonometric identity.

2 If $y = \sin^2 x$, show that $\dfrac{dy}{dx} = \sin 2x$.

3 Differentiate with respect to x:

(a) $\ln(x^2 + 3)$ (b) $\ln(2x - 4)^5$ (c) $\ln(\cos 2x)$ (d) $\ln(6x - 1)$

4 Find $\dfrac{dy}{dx}$, simplifying your answers.

(a) $y = x^2 \cos 3x$ (b) $y = e^{2x} \sin x$ (c) $y = x \ln x$ (d) $y = (x + 1)(2x - 3)^4$

5 Find the equation of the tangent to the curve $y = x\sqrt{1 + 2x}$ at the point $(4, 12)$.

6 Use the quotient rule to find $\dfrac{dy}{dx}$, simplifying your answers where necessary.

(a) $y = \dfrac{\sin 2x}{x}$ when $t = 2$ (b) $y = \dfrac{2x - 1}{3x + 4}$ (c) $y = \dfrac{\ln(x + 2)}{x + 2}$

7 A curve is defined parametrically by $x = t^2$, $y = t^3$.

(a) Find $\dfrac{dy}{dx}$ when $t = 2$.

(b) Find the Cartesian equation of the curve.

8 A curve has equation $x^2 + xy^2 - 5 = 0$.

(a) Find the value of $\dfrac{dy}{dx}$ at $(1, 2)$.

(b) Find the equation of the normal when $x = 1$, in the form $ax + by + c = 0$.

8 (a) -1.5 (b) $2x - 3y + 4 = 0$

7 (a) 3 (b) $x^3 = y^2$

6 (a) $\dfrac{2x \cos 2x - \sin 2x}{x^2}$ (b) $\dfrac{11}{(3x + 4)^2}$ (c) $\dfrac{1 - \ln(x + 2)}{(x + 2)^2}$

5 $3y = 13x - 16$

4 (a) $x(2 \cos 3x - 3x \sin 3x)$ (b) $e^{2x}(\cos x + 2 \sin x)$ (c) $1 + \ln x$ (d) $5(2x + 1)(2x - 3)^3$

3 (a) $\dfrac{2x}{x^2 + 3}$ (b) $\dfrac{5}{x - 2}$ (c) $-2 \tan 2x$ (d) $\dfrac{6}{6x - 1}$

1 (a) $12 \cos 6x$ (b) $-12 \sin x \cos^2 x$ (c) $6 \sec^2 3x$ (d) $\cos x e^{\sin x}$ (e) $3x^2 \cos(x^3)$

1.6 Integration

After studying this section you should be able to:

- integrate e^x and simple variations of this function with respect to x
- integrate trigonometrical functions
- recognise an integral that leads to a logarithmic function
- integrate using partial fractions
- integrate using a substitution
- use integration by parts to integrate a product
- find the area under a curve defined parametrically
- use integration to find a volume of revolution
- solve first order differential equations when the variables can be separated
- understand exponential growth and decay

LEARNING SUMMARY

The exponential function e^x

AQA	C3
EDEXCEL	C4
OCR	C3
WJEC	C3
CCEA	C3

The fact that differentiation leaves the exponential function unchanged was stated on page 36. Since integration is the reverse process of differentiation it follows that

$$\int e^x \, dx = e^x + c \quad \text{and} \quad \int a e^x \, dx = a \int e^x \, dx = a e^x + c.$$

KEY POINT

An extension of this result, using the reverse of the chain rule, is

$$\int e^{ax+b} \, dx = \frac{1}{a} e^{ax+b} + c.$$

KEY POINT

Examples

1 $\int 4e^x \, dx = 4e^x + c$

2 $\int \tfrac{1}{2} e^{3x-1} \, dx = \tfrac{1}{6} e^{3x-1} + c$

Trigonometric functions

AQA	C3
EDEXCEL	C4
OCR	C4
WJEC	C3
CCEA	C3

The following integrals are obtained by applying the reverse process of differentiating the three main trigonometric functions (page 38).

You should learn these and practise using them.

$$\int \cos x \, dx = \sin x + c \qquad \int \cos(ax+b) \, dx = \frac{1}{a} \sin(ax+b) + c$$

$$\int \sin x \, dx = -\cos x + c \qquad \int \sin(ax+b) \, dx = -\frac{1}{a} \cos(ax+b) + c$$

$$\int \sec^2 x \, dx = \tan x + c \qquad \int \sec^2(ax+b) \, dx = \frac{1}{a} \tan(ax+b) + c$$

Example

It is a good idea to check your answer by differentiating.

$$\int (4 \cos 2x - \sin 2x) dx = 2 \sin 2x + \tfrac{1}{2} \cos 2x + c$$

Example

When integrating, the angle must be in radians.

$$\int_{\frac{\pi}{3}}^{\frac{\pi}{2}} \sec^2(\tfrac{1}{2}x)\,dx = \left[\frac{1}{\frac{1}{2}}\tan(\tfrac{1}{2}x)\right]_{\frac{\pi}{3}}^{\frac{\pi}{2}}$$

$$= 2\left[\tan\left(\tfrac{1}{2}x\right)\right]_{\frac{\pi}{3}}^{\frac{\pi}{2}}$$

$$= 2\left(\tan\frac{\pi}{4} - \tan\frac{\pi}{6}\right)$$

Learn the exact values of the trigonometric ratios of $\frac{\pi}{6}, \frac{\pi}{4}, \frac{\pi}{3}$ (30°, 45°, 60°).

$$= 2\left(1 - \frac{1}{\sqrt{3}}\right)$$

These results are very useful and can be recognised easily:

$$\frac{d}{dx}(\sin^{n+1}x) = (n+1)\sin^n x \cos x \quad \Rightarrow \quad \int \sin^n x \cos x\,dx = \frac{1}{n+1}\sin^{n+1}x + c$$

$$\frac{d}{dx}(\cos^{n+1}x) = -(n+1)\cos^n x \sin x \quad \Rightarrow \quad \int \cos^n x \sin x\,dx = -\frac{1}{n+1}\cos^{n+1}x + c$$

Examples

$$\int \sin^5 x \cos x\,dx = \tfrac{1}{6}\sin^6 x + c$$

$$\int \sin x \cos^3 x\,dx = -\tfrac{1}{4}\cos^4 x + c$$

Odd powers of sin x and cos x

Example

Evaluate $\displaystyle\int_0^{\frac{\pi}{6}} \cos^3 x\,dx$

$$\int_0^{\frac{\pi}{6}} \cos^3 x\,dx = \int_0^{\frac{\pi}{6}} \cos x(\cos^2 x)\,dx$$

Use $\cos^2 x = 1 - \sin^2 x$.

$$= \int_0^{\frac{\pi}{6}} \cos x(1 - \sin^2 x)\,dx$$

$$= \int_0^{\frac{\pi}{6}} (\cos x - \cos x \sin^2 x)\,dx$$

Recognise the integral of $\cos x \sin^2 x$.

$$= \left[\sin x - \tfrac{1}{3}\sin^3 x\right]_0^{\frac{\pi}{6}}$$

$$= \tfrac{1}{2} - \tfrac{1}{24} = \tfrac{11}{24}$$

Even powers of cos x and sin x

To integrate even powers of sin x and cos x, use the double angle identities for $\cos 2x$.

$$\cos 2x = 2\cos^2 x - 1 \quad \Rightarrow \quad \cos^2 x = \tfrac{1}{2}(1 + \cos 2x)$$

$$\cos 2x = 1 - 2\sin^2 x \quad \Rightarrow \quad \sin^2 x = \tfrac{1}{2}(1 - \cos 2x)$$

Example

$$\int \cos^2 x \, dx = \tfrac{1}{2} \int (1 + \cos 2x) \, dx$$

$$= \tfrac{1}{2}(x + \tfrac{1}{2} \sin 2x) + c$$

To integrate $\cos^4 x$, write it as $(\cos^2 x)^2 = (\tfrac{1}{2}(1 + \cos 2x))^2$.

The reciprocal function $\dfrac{1}{x}$

AQA	C3
EDEXCEL	C4
OCR	C3
WJEC	C3
CCEA	C3

The result $\displaystyle\int x^n \, dx = \dfrac{x^{n+1}}{n+1} + c$ holds for any value of n other than $n = -1$.

In this special case, the result takes a different form:

Remember that $x^{-1} = \dfrac{1}{x}$.

> **KEY POINT**
>
> $$\int \frac{1}{x} \, dx = \ln|x| + c.$$

This extends to $\displaystyle\int \frac{k}{x} \, dx = k \int \frac{1}{x} \, dx = k \ln|x| + c,$

and to $\displaystyle\int \frac{1}{kx} \, dx = \frac{1}{k} \int \frac{1}{x} \, dx = \frac{1}{k} \ln|x| + c.$ (where k is a constant)

For example, $\displaystyle\int \frac{2}{x} \, dx = 2 \ln|x| + c$ and $\displaystyle\int \frac{1}{3x} \, dx = \frac{1}{3} \ln|x| + c.$

Integrals leading to a logarithmic function

AQA	C3
EDEXCEL	C4
OCR	C4
WJEC	C4
CCEA	C4

Remember that $\dfrac{d}{dx} \ln(f(x)) = \dfrac{f'(x)}{f(x)}.$

Applying the process in reverse gives:

> **KEY POINT**
>
> $$\int \frac{f'(x)}{f(x)} \, dx = \ln|f(x)| + c.$$

Examples

(a) $\displaystyle\int \frac{x^2}{2x^3 - 2} \, dx = \tfrac{1}{6} \ln|2x^3 - 2| + c$

    $\dfrac{d}{dx}(2x^3 - 2) = 6x^2$ so a factor of $\tfrac{1}{6}$ is needed.

(b) $\displaystyle\int \tan x \, dx = \int \frac{\sin x}{\cos x} \, dx$

$$= -\ln|\cos x| + c$$

$$= \ln|\sec x| + c$$

    $-\ln a = \ln(a^{-1}).$

Care must be taken with definite integrals of this type.
These can be evaluated between the limits a and b only if $f(x)$ exists for all values between a and b. Note that $f(a)$ and $f(b)$ will have the same sign.

$$\int_a^b \frac{f'(x)}{f(x)} \, dx = \left[\ln|f(x)| \right]_a^b = (\ln|f(b)| - \ln|f(a)|)$$

Key points from AS

- **Area under curve**
  Revise AS page 57

Example

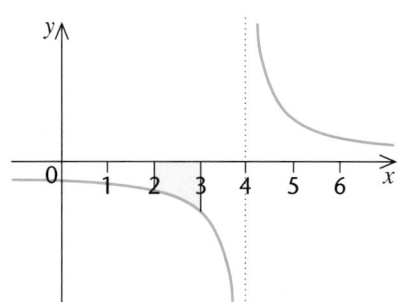

The diagram shows the curve $y = \dfrac{1}{x-4}$.

Find the area between the curve, the x-axis, and the lines $x = 2$ and $x = 3$.

$$\int_a^b y\,\mathrm{d}x = \int_2^3 \frac{1}{x-4}\,\mathrm{d}x$$

$$= \Big[\ln|x-4|\Big]_2^3$$

$\ln|-2| = \ln 2$
$\ln 1 = 0$.

$$= \ln|-1| - \ln|-2|$$
$$= \ln 1 - \ln 2$$
$$= -\ln 2\,(= -0.69\ \ldots)$$

Note that the formula gives a negative value, confirming that the area is below the x-axis.

Shaded area $= \ln 2\,(= 0.69\ \ldots)$ square units.

$f(3)$ and $f(5)$ do not have the same sign.

Note that $\displaystyle\int_3^5 \frac{1}{x-4}\,\mathrm{d}x$ cannot be evaluated, as the curve is undefined at $x = 4$.

Partial fractions

AQA	C4
EDEXCEL	C4
OCR	C4
WJEC	C4
CCEA	C4

Some rational functions can be integrated by expressing them in partial fraction form.

Example

Find $\displaystyle\int \frac{x+7}{(x-2)(x+1)}\,\mathrm{d}x$, simplifying your answer.

Decompose into partial fractions (see p. 19).

$$\frac{x+7}{(x-2)(x+1)} \equiv \frac{3}{x-2} - \frac{2}{x+1}$$

The working for this example is shown on p.19.

$$\Rightarrow \int \frac{x+7}{(x-2)(x+1)}\,\mathrm{d}x = \int \left(\frac{3}{x-2} - \frac{2}{x+1}\right)\mathrm{d}x$$

$$= 3\ln(x-2) - 2\ln(x+1) + c$$

$$= \ln k\,\frac{(x-2)^3}{(x+1)^2}$$

$\log a - \log b = \log\left(\dfrac{a}{b}\right)$

$\log a^n = n \log a$.

Key points from AS

- **Logarithms**
  Revise AS page 31

Note: Sometimes the constant c is incorporated into the log function by letting $c = \ln k$.

Substitution

AQA	C3
EDEXCEL	C4
OCR	C3/C4
WJEC	C4
CCEA	C4

Sometimes an integral is made easier by using a substitution.

This transforms the integral with respect to one variable, say x, into an integral with respect to a related variable, say u.

$$\int f(x)\mathrm{d}x = \int f(x)\frac{\mathrm{d}x}{\mathrm{d}u}\,\mathrm{d}u.$$

KEY POINT

This method is known as the method of **substitution** or **change of variable**.

> This could also be done by using the substitution $u = 2x + 1$.

Example

Find $\displaystyle\int x\sqrt{2x+1}\ \mathrm{d}x$, using the substitution $u = \sqrt{2x+1}$.

$$\int x\sqrt{2x+1}\ \mathrm{d}x = \int x\sqrt{2x+1}\ \frac{\mathrm{d}x}{\mathrm{d}u}\,\mathrm{d}u \qquad\qquad \text{Side working}$$

> Show the side working as part of your answer.

$$= \int \tfrac{1}{2}(u^2 - 1)u \times u\,\mathrm{d}u \qquad\qquad u = \sqrt{2x+1}$$

$$= \frac{1}{2}\int (u^4 - u^2)\mathrm{d}u \qquad\qquad u^2 = 2x+1 \implies x = \tfrac{1}{2}(u^2-1)$$

> When simplifying, take out factors as soon as possible.

$$= \frac{1}{2}\left(\frac{u^5}{5} - \frac{u^3}{3}\right) + c \qquad\qquad \frac{\mathrm{d}x}{\mathrm{d}u} = u$$

$$= \frac{u^3}{2}\left(\frac{u^2}{5} - \frac{1}{3}\right) + c$$

> Write the answer in terms of x, simplifying as much as possible.

$$= \tfrac{1}{30}\,u^3(3u^2 - 5) + c \qquad\qquad 3u^2 - 5 = 3(2x+1) - 5$$

$$= \tfrac{1}{30}(2x+1)^{\frac{3}{2}}(6x - 2) + c \qquad\qquad = 6x - 2$$

$$= \tfrac{1}{15}(2x+1)^{\frac{3}{2}}(3x - 1) + c \qquad\qquad = 2(3x - 1)$$

When evaluating definite integrals using the method of substitution, change the x limits to u limits and substitute them into the working as soon as possible. There is no need to tidy up the expression first.

Sometimes it is possible to bypass the method of substitution by **recognising** an integral as the derivative of a particular function.

Examples

> This type of integral can always be done by using a substitution.

(a) $\displaystyle\int (2x+1)^5\ \mathrm{d}x = \tfrac{1}{12}(2x+1)^6 + c$

> Recognise that
> $\dfrac{\mathrm{d}}{\mathrm{d}x}(2x+1)^6 = 12(2x+1)^5.$

(b) $\displaystyle\int x\sqrt{2x^2+1}\ \mathrm{d}x = \tfrac{1}{6}(2x^2+1)^{\frac{3}{2}} + c$

> Recognise that
> $\dfrac{\mathrm{d}}{\mathrm{d}x}(2x^2+1)^{\frac{3}{2}} = 6x(2x^2+1)^{\frac{1}{2}}.$

General result:

$$\int f'(x)[f(x)]^n\ \mathrm{d}x = \frac{1}{n+1}[f(x)]^{n+1} + c$$

Integration by parts

AQA	C3
EDEXCEL	C4
OCR	C4
WJEC	C4
CCEA	C4

It is sometimes possible to integrate a product using integration by parts.

> **KEY POINT**
>
> If u and v are functions of x, then $\int u \dfrac{dv}{dx} dx = uv - \int v \dfrac{du}{dx} dx$.

Before using integration by parts, check whether the function can be simplified, or whether the integral can be recognised directly or done using a substitution.

Example

Find $\int 2x \sin x \, dx$.

Let $u = 2x$, then $\dfrac{du}{dx} = 2$

Let $\dfrac{dv}{dx} = \sin x$ then $v = -\cos x$

$\therefore \quad \int 2x \sin x \, dx = 2x(-\cos x) - \int (-\cos x \times 2) dx$

$\qquad\qquad = -2x \cos x + \int 2 \cos x \, dx$

$\qquad\qquad = -2x \cos x + 2 \sin x + c$

> Integration by parts can only be used if v can be found from $\dfrac{dv}{dx}$ and it is possible to find $\int v \dfrac{du}{dx} dx$.

Sometimes the integration by parts process has to be carried out more than once. This happens when integrating expressions such as $x^2 e^x$ or $x^2 \cos x$.

Definite integrals

$$\int_a^b u \frac{dv}{dx} dx = \left[uv \right]_a^b - \int_a^b v \frac{du}{dx} dx$$

Example

$\displaystyle\int_0^1 x e^{2x} \, dx = [x \tfrac{1}{2} e^{2x}]_0^1 - \int_0^1 \tfrac{1}{2} e^{2x} \, dx$

$\qquad\qquad = \tfrac{1}{2} e^2 - [\tfrac{1}{4} e^{2x}]_0^1$

$\qquad\qquad = \tfrac{1}{2} e^2 - (\tfrac{1}{4} e^2 - \tfrac{1}{4}) = \tfrac{1}{4} e^2 + \tfrac{1}{4}$

Working:

Let $u = x$, then $\dfrac{du}{dx} = 1$

Let $\dfrac{dv}{dx} = e^{2x}$, then $v = \tfrac{1}{2} e^{2x}$

Some functions containing $\ln x$ can be integrated using integration by parts. A useful strategy is to take $\ln x$ as u.

> Often a polynomial function is taken as u because it becomes of lower degree when differentiated. Integrals involving $\ln x$ are an exception, since $\ln x$ is not easy to integrate, but can be differentiated.

Example

$\displaystyle\int x^2 \ln x \, dx = \int (\ln x \times x^2) dx$

$\qquad = \ln x \times \left(\dfrac{x^3}{3}\right) - \int \left(\dfrac{x^3}{3} \times \dfrac{1}{x}\right) dx$

$\qquad = \tfrac{1}{3} x^3 \ln x - \tfrac{1}{3} \int x^2 \, dx$

$\qquad = \tfrac{1}{3} x^3 \ln x - \tfrac{1}{9} x^3 + c$

Working:

Let $u = \ln x$, then $\dfrac{du}{dx} = \dfrac{1}{x}$

Let $\dfrac{dv}{dx} = x^2$, then $v = \dfrac{x^3}{3}$

This method is particularly useful when integrating ln x.
Think of it as integrating the product ln $x \times 1$ as follows:

$$\int (\ln x \times 1)\,dx = \ln x \times x - \int \left(x \times \frac{1}{x}\right)dx$$

$$= x \ln x - \int 1\,dx$$

$$= x \ln x - x + c$$

Working:

Let $u = \ln x$, then $\dfrac{du}{dx} = \dfrac{1}{x}$

Let $\dfrac{dv}{dx} = 1$, then $v = x$

Parametric integration

EDEXCEL   C4

When a curve is defined in terms of a parameter such that $x = f(t)$ and $y = g(t)$ then the formula for the area under a curve can be adapted as follows:

$$A = \int_{x_1}^{x_2} y\,dx = \int_{t_1}^{t_2} g(t)\,\frac{dx}{dt}\,dt.$$

Example
Using parametric integration, find the area of the ellipse defined by $x = 2 \cos \theta$, $y = 3 \sin \theta$ ($0 \leqslant \theta \leqslant 2\pi$).

The sign of the area given by the formula depends on the direction in which the curve is being traced out as the parameter increases: positive for clockwise, negative for anticlockwise.

In general, area of ellipse defined by $x = a \cos \theta$, $y = b \sin \theta$ is πab.

$$A = \int_0^{2\pi} y\,\frac{dx}{d\theta}\,d\theta$$

$$= \int_0^{2\pi} 3 \sin \theta(-2 \sin \theta)d\theta$$

$$= -6\int_0^{2\pi} \sin^2 \theta\,d\theta$$

$$= -6\int_0^{2\pi} \tfrac{1}{2}(1 - \cos 2\theta)d\theta$$

$$= -3\left[\theta - \tfrac{1}{2}\sin 2\theta\right]_0^{2\pi}$$

$$= -3(2\pi - 0 - (0 - 0))$$

$$= -6\pi$$

This curve is traced out in an anticlockwise direction

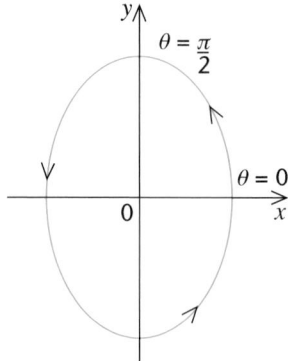

The area of the ellipse is 6π square units.

Volume of revolution

About x-axis

AQA   C3
EDEXCEL   C4
OCR   C3
WJEC   C4
CCEA   C4

The shaded region R is bounded by the curve $y = f(x)$, the x-axis and the lines $x = a$ and $x = b$.

Rotating R completely about the x-axis forms a solid figure. The volume of this figure is called the volume of revolution and is given by

$$V = \int_a^b \pi y^2\,dx$$

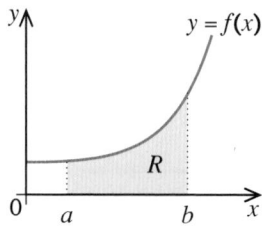

The corresponding result for rotation about the y-axis is

$$\int_a^b \pi x^2 \, dy$$

where the region is bounded by the curve $y = f(x)$, the y-axis and the lines $y = a$ and $y = b$.

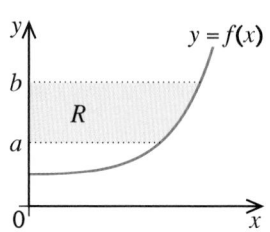

Example Find a formula for the volume of a cone of height h and base radius r.

The volume of the cone is given by the volume of revolution of the shaded region shown.

The straight line has gradient $\dfrac{r}{h}$ and passes through the origin, so its equation is $y = \dfrac{r}{h} x$.

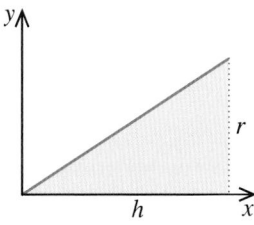

Using the formula for volume of revolution gives:

$$V = \int_0^h \pi \left(\frac{r}{h} x\right)^2 dx = \frac{\pi r^2}{h^2} \int_0^h x^2 \, dx$$

$$= \frac{\pi r^2}{h^2} \left[\frac{x^3}{3}\right]_0^h$$

$$= \frac{\pi r^2}{h^2} \times \frac{h^3}{3}$$

$$= \frac{\pi r^2 h}{3}.$$

Differential equations

A **first order differential equation** in x and y contains only the first differential coefficient.

Examples $\dfrac{dy}{dx} = -2x,$ $\dfrac{dy}{dx} = \dfrac{x}{y},$ $y\dfrac{dy}{dx} = 3$

If $\dfrac{dy}{dx} = -2x$, then integrating with respect to x gives $y = -x^2 + c$.

> The general solution contains an arbitrary integration constant.

This is the **general solution** of the differential equation and it can be illustrated by a family of curves.

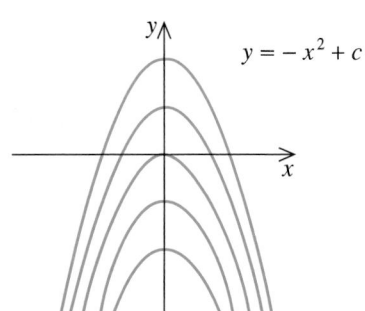

A **particular solution** is a specific member of the family and it can be found from additional information.

The value of c is found to give a particular solution.

For example, if the curve $y = -x^2 + c$ goes through the point $(2, 0)$, then $0 = -(2)^2 + c \Rightarrow c = 4$. The particular solution is $y = -x^2 + 4$.

Separating the variables

A differential equation that can be written in the form $f(y) \dfrac{dy}{dx} = g(x)$

can be solved by **separating the variables**, where $\displaystyle\int f(y)dy = \int g(x)dx$.

Example

Solve (a) $\dfrac{dy}{dx} = \dfrac{x}{y}$ (b) $\dfrac{dy}{dx} = y$

(a) $\dfrac{dy}{dx} = \dfrac{x}{y}$

Separate the variables to get the format $\int \dots dy = \int \dots dx.$

$$\int y \, dy = \int x \, dx \Rightarrow \tfrac{1}{2}y^2 = \tfrac{1}{2}x^2 + c$$

(b) $\dfrac{dy}{dx} = y$

This is a special case, when $g(x) = 1$.

$$\dfrac{1}{y}\dfrac{dy}{dx} = 1$$

$$\int \dfrac{1}{y} \, dy = \int 1 \, dx$$

The answer can be left like this.

$$\ln y = x + c$$
$$y = e^{x+c} = e^x \times e^c = Ae^x \quad (\text{where } A = e^c)$$

So $\dfrac{dy}{dx} = y \Rightarrow y = Ae^x$

The letter A is often used here, but it could be any letter (other than e, x or y, of course!).

Exponential growth and decay

AQA	C4
EDEXCEL	C4
OCR	C3.C4
WJEC	C4
CCEA	C3

You may have to form a differential equation from given information. Often this models exponential growth or exponential decay.

Exponential growth   **Exponential decay**

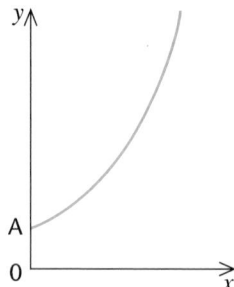 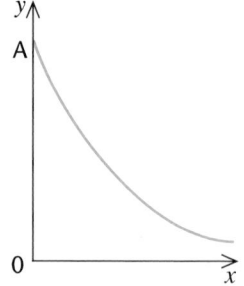

In exponential decay, the fact that y is decreasing with respect to x is usually shown by the negative, keeping $k > 0$.

$$\dfrac{dy}{dx} = ky \Rightarrow y = Ae^{kx} \quad (k > 0) \qquad\qquad \dfrac{dy}{dx} = -ky \Rightarrow y = Ae^{-kx} \quad (k > 0)$$

An **example** of exponential growth is the growth of a population where

p_0 is the initial population.

$$\dfrac{dp}{dt} = kt \Rightarrow p = p_0 e^{-kt}$$

Examples of exponential decay are:

> m_0 is the initial mass.

(a) the disintegration of radioactive materials, $\dfrac{dm}{dt} = -km \Rightarrow m = m_0 e^{-kt}$

> T_s is temperature of surroundings, T_0 is initial temperature of the object.

(b) Newton's Law of Cooling, $\dfrac{dT}{dt} = -k(T - T_s) \Rightarrow T - T_s = (T_0 - T_s)e^{-kt}$

Progress check

1 Integrate with respect to x.

(a) $\dfrac{3}{x}$ (b) $\dfrac{x+3}{x^2}$ (c) $3e^x - \dfrac{1}{x}$ (d) $3e^{4-x}$

2 The region bounded by the curve $y = e^{\frac{x}{2}}$, the lines $x = 1$ and $x = 3$ and the x-axis is rotated completely about the x-axis. Find the volume of revolution.

3 (a) $\displaystyle\int 4\sin 2x\, dx$ (b) $\displaystyle\int \sec^2 2x\, dx$ (c) $\displaystyle\int \cos\left(\tfrac{1}{2}x\right)dx$

(d) $\displaystyle\int 4\sin^2 x\, dx$ (e) $\displaystyle\int \cos^2 2x\, dx$ (f) Evaluate $\displaystyle\int_0^{\frac{1}{6}\pi} \sin^4 x \cos x\, dx$

4 Find (a) $\displaystyle\int \dfrac{1}{2x+5}\, dx$ (b) $\displaystyle\int \cot x\, dx$ (c) $\displaystyle\int \dfrac{3x+2}{x^2-x-12}\, dx$

> Split (c) into partial fractions first.

> Remember that $\dfrac{dx}{du} = \dfrac{1}{du/dx}$.

5 Using the substitution $u = 2 + e^{2x}$, or otherwise, find $\displaystyle\int \dfrac{e^{2x}}{2+e^{2x}}\, dx$

6 (a) Find (i) $\displaystyle\int xe^{-x}\, dx$ (ii) $\displaystyle\int x^3 \ln x\, dx$ (b) Evaluate $\displaystyle\int_0^{\frac{1}{2}\pi} x\cos x\, dx$

7 Solve the differential equation $\dfrac{dy}{dx} = 2xy$, given that $y = 3$ when $x = 0$.

1.7 Numerical methods

After studying this section you should be able to:

- locate an interval containing a root of an equation by finding a change of sign
- use simple iteration to solve equations
- find an approximate solution to an equation using the Newton-Raphson method
- find the approximate value of an integral using the trapezium rule, Simpson's rule, the mid-ordinate rule

LEARNING SUMMARY

Change of sign

AQA	C3
EDEXCEL	C3
OCR	C3
WJEC	C3
CCEA	C3

If there is an interval from $x = a$ to $x = b$ in which the graph of a function has no breaks then the function is said to be **continuous** on the interval.

All polynomial functions are continuous everywhere.

For example, the graph of $y = x^2 - 3$ is continuous everywhere.

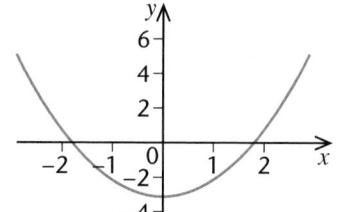

However, the graph of $y = \dfrac{1}{x - 3}$ is not continuous on any interval that contains the value $x = 3$.

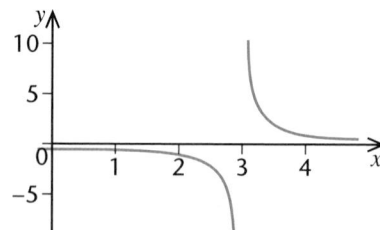

> If $f(x)$ is continuous between $x = a$ and $x = b$ and if $f(a)$ and $f(b)$ have different signs, then a root of $f(x) = 0$ lies in the interval from a to b.
>
> **KEY POINT**

Example Given that $f(x) = e^x - 10x$, show that the equation $f(x) = 0$ has a root between 3 and 4.

$f(3) = -9.914\ldots < 0$

$f(4) = 14.598\ldots > 0.$

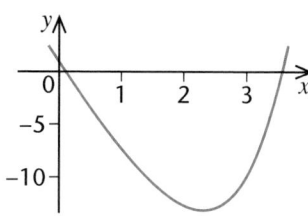

The change of sign shows that $f(x) = 0$ has a root between 3 and 4.

You can continue this process, called a **decimal search**, to trap the root in a smaller and smaller interval. The table gives the values of the function in steps of 0.1.

Most graphic calculators can tabulate values of a function and this gives a very efficient way to obtain the information.

x	3.1	3.2	3.3	3.4	3.5	3.6	3.7	3.8	3.9
$f(x)$	−8.802	−7.467	−5.887	−4.035	−1.884	0.5982	3.4473	6.7011	10.402

This shows that the root lies between 3.5 and 3.6
At the mid-point $f(3.55) = -0.6866\ldots < 0$ and so the root lies between 3.55 and 3.6.

This is a good way to establish the value of a root to a particular level of accuracy.

3.5	3.55	3.6

It follows that the value of the root is 3.6 to 1 d.p.

Iteration

AQA C3
EDEXCEL C3
OCR C3
WJEC C3
NICCEA C3

There are different methods for producing an iterative formula. You only need to know about simple iteration for this component.

To solve an equation by **iteration** you start with some approximation to a root and improve its accuracy by substituting it into a formula. You can then repeat the process until you have the desired level of accuracy.

Using **simple iteration**, the first step is to rearrange the equation to express x as a function of itself. This function defines the iterative formula that you need.

For example, the equation $e^x - 10x = 0$ can be rearranged as

$$e^x = 10x$$

$$\Rightarrow x = \ln(10x).$$

This may now be turned into an iterative formula for finding x.

$$x_{n+1} = \ln(10x_n).$$

Starting with $x_1 = 3$, $x_2 = \ln 30 = 3.401 \ldots$ and so on. This produces a sequence of values:

The values may be found very easily using a calculator's Ans function:
Key in 3 =
followed by
ln(10Ans) = = =
to produce the sequence.
(Depending on the make of your calculator you may need to use EXE or ENTER in place of = .)

$x_2 = 3.401 \ldots$ $x_6 = 3.5760 \ldots$
$x_3 = 3.526 \ldots$ $x_7 = 3.5768 \ldots$
$x_4 = 3.563 \ldots$ $x_8 = 3.5770 \ldots$
$x_5 = 3.573 \ldots$ $x_9 = 3.5771 \ldots$

The calculator display soon settles on 3.577152064. The value of the root is $x = 3.577$ to 4 s.f.

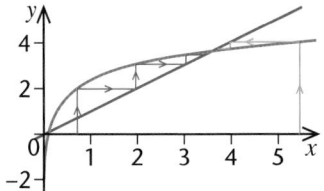

Start from a point on the x-axis, move up to the curve and across to $y = x$. Then move up to the curve and across to $y = x$ again. Continue in the same way. Each of these stages corresponds to one iteration of the formula.

The diagram shows how the process converges from starting point on either side of the root.

A different rearrangement of the original equation gives $x = \dfrac{e^x}{10}$ and so the corresponding iterative formula is $x_{n+1} = \dfrac{e^{x_n}}{10}$.

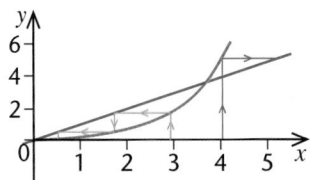

Notice how the movement is away from the upper root this time. If the starting value is smaller than the upper root then the process converges to the lower root.
A starting value above the upper root fails to converge to either root.

You may need to try several rearrangements in order to converge to a particular root of an equation by this method.

Starting with $x = 3$, this rearrangement fails to converge to the root between 3 and 4 but it does co verge to the other root of the equation.

The value of this root is $x = 0.1118$ to 4 s.f.

In some exam questions you may be given the iteration formula to start with.

Example Starting with $x_1 = 1$, use the formula $x_{n+1} = \dfrac{x_n^2 - 5x_n}{10} - 3$ to find $x_2, x_3, \ldots, x_8$ and describe the long term behaviour of the sequence.

Try the key sequence:
1 =
$(\text{Ans}^2 - 5\text{Ans})/10 - 3$
=
=
=
=

$x_2 = -3.4$

$x_3 = -0.144$

$x_4 = -2.9259264$

$x_5 = -0.6809322$

$x_6 = -2.6131669$

$x_7 = -1.0105523$

$x_8 = -2.3926032$

The sequence oscillates but converges to -1.787 to 4 s.f. which is the lower root of the equation

$$x = \frac{x^3 - 5x}{10} - 3,$$

which simplifies to

$$x^2 - 15x - 30 = 0.$$

The oscillations correspond to a cobweb pattern on the diagram.

Newton-Raphson method

CCEA ▸ C3

The **Newton-Raphson method** can be used to find an approximate solution to an equation.

Consider the equation $f(x) = 0$, with unknown root α, i.e. $f(\alpha) = 0$.

This process can then be repeated several times to obtain a more accurate value for the solution.

> **KEY POINT**
>
> If x_n is a good approximation to α, then a better approximation* is given by
>
> $$x_{n+1} = x_n - \frac{f(x_n)}{f'(x_n)}.$$

* There are cases when this method fails to converge, for example when $f'(\alpha)$ is near to zero or when $f'(\alpha)$ is very large. There may also be problems if there are discontinuities in the curve.

Example

The equation $f(x) = 2\cos x - x^2$ has root α.

(a) Show that α lies between 1 and 1.1.

(b) Using 1 as a first approximation, find α correct to 2 decimal places.

$$f(x) = 2\cos x - x^2$$

$$f(1) = 2\cos 1 - 1 = 0.0806$$

$$f(1.1) = 2\cos 1.1 - 1.1^2 = -0.3028\ldots$$

Remember to work in radians.

Since $f(1) > 0$ and $f(1.1) < 0$, there must be a value between $x = 1$ and $x = 1.1$ for which $f(x) = 0$. Therefore α lies between 1 and 1.1.

$$f'(x) = -2\sin x - 2x$$

Taking $x_1 = 1$,

$$x_2 = 1 - \frac{f(1)}{f'(1)}$$

$$= 1 - \frac{2\cos 1 - 1}{-2\sin 1 - 2}$$

$$= 1 - \frac{0.0806\ldots}{-3.6829\ldots}$$

$$= 1.021885\ldots$$

Taking $x_2 = 1.02189$,

$$x_3 = 1.02189 - \frac{f(1.02189)}{f'(1.02189)}$$

$$= 1.02189 - \frac{-0.000750\ldots}{-3.7499\ldots}$$

$$= 1.02169\ldots$$

The approximations agree to 2 d.p.

$$\therefore \qquad \alpha = 1.02 \ (2 \text{ d.p.})$$

Numerical integration

AQA	C2
EDEXCEL	C2/C4
OCR	C2
WJEC	C2

The trapezium rule

The value of $A = \displaystyle\int_a^b f(x)\,dx$ represents the area under the graph of $y = f(x)$ between $x = a$ and $x = b$. You can find an approximation to this value using the **trapezium rule**. This is especially useful if the function is difficult to integrate.

The area under the curve between a and b may be divided into n strips of equal width d.

It follows that $d = \dfrac{b-a}{n}$.

Each strip is approximately a trapezium and so the total area is approximately

$$\tfrac{1}{2}d(y_0 + y_1) + \tfrac{1}{2}d(y_1 + y_2) + \tfrac{1}{2}d(y_2 + y_3) + \ldots + \tfrac{1}{2}d(y_{n-2} + y_{n-1}) + \tfrac{1}{2}d(y_{n-1} + y_n).$$

This simplifies to give the formula known as the trapezium rule:

> $$\int_a^b f(x)\,dx \approx \tfrac{1}{2}d\left(y_0 + 2(y_1 + y_2 + y_3 + \ldots y_{n-1}) + y_n\right) \text{ where } d = \dfrac{b-a}{n}.$$
>
> **KEY POINT**

Example

Use the trapezium rule with five strips to estimate the value of $\displaystyle\int_1^2 2^x\,dx$.

It's useful to tabulate the information:

Work to a greater level of accuracy in your table than you intend to give in your final answer.

x	1	1.2	1.4	1.6	1.8	2
2^x	2	2.2974	2.6390	3.0314	3.4822	4
	y_0	y_1	y_2	y_3	y_4	y_5

In this case, the working was done to 5 s.f. and the final answer is given to 3 s.f.

A graphic calculator with a numerical integration function gives the value of the area as 2.8853901. This shows that no more than 3 s.f. could be justified.

$$d = \frac{2-1}{5} = 0.2$$

$$\int_1^2 2^x\,dx \approx \tfrac{1}{2} \times 0.2 \times (2 + 2(2.2974 + 2.6390 + 3.0314 + 3.4822) + 4)$$

$$= 2.89 \quad \text{to} \quad 3 \text{ s.f.}$$

Simpson's rule

AQA	C3
OCR	C3
WJEC	C2
CCEA	C3

This method divides the area into an **even number** of parallel strips and approximates the areas of pairs of strips using the following formula:

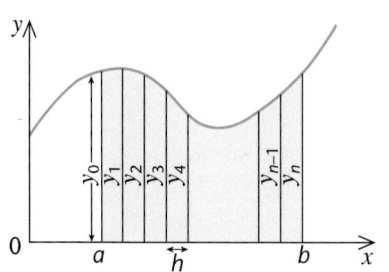

$$\int_a^b f(x)\,\mathrm{d}x \approx \tfrac{1}{3}h\{(y_0 + y_n) + 4(y_1 + y_3 + \ldots + y_{n-1}) + 2(y_2 + y_4 + \ldots + y_{n-2})\}$$

where $h = \dfrac{b-a}{n}$ and n is even.

Note that the first ordinate has been labelled y_0 and the last y_n. There are n strips and $n+1$ ordinates.

Example

Estimate $\displaystyle\int_0^1 e^{x^2}\,\mathrm{d}x$ using Simpson's rule with 10 strips.

Tabulating the results helps in doing the final calculation.

If you have multiple memories in your calculator, retain the figures. Otherwise approximate say to 4 decimal places.

x	y	First and last ordinates	Odd ordinates	Other ordinates
0	y_0	1		
0.1	y_1		1.010 …	
0.2	y_2			1.040 …
0.3	y_3		1.094 …	
0.4	y_4			1.173 …
0.5	y_5		1.284 …	
0.6	y_6			1.433 …
0.7	y_7		1.632 …	
0.8	y_8			1.896 …
0.9	y_9		2.247 …	
1	y_{10}	2.718 …		
Totals		3.718 …	7.268 …	5.544 …

$h = \dfrac{1-0}{10} = 0.1.$

$$\int_a^b f(x)\,\mathrm{d}x \approx \tfrac{1}{3}h\{(y_0 + y_{10}) + 4(y_1 + y_3 + y_5 + y_7 + y_9) + 2(y_2 + y_4 + y_6 + y_8)\}$$

$$= \tfrac{1}{3} \times 0.1 \times \{3.718\ldots + 4(7.268\ldots) + 2(5.544\ldots)\}$$

$$= 1.463 \text{ (3 s.f.)}$$

AQA ▷ C3

The mid-ordinate rule

The more rectangles that are taken, the better the approximation.

The area is split into rectangles of equal width h.
The height of each rectangle is taken as the y-value halfway along each rectangle.

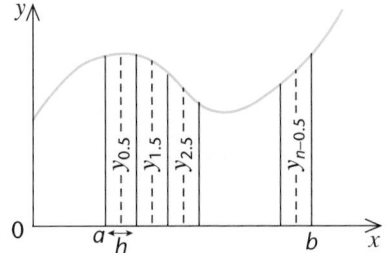

Denoting the mid-ordinates by $y_{0.5}, y_{1.5}, \ldots, y_{n-0.5}$, then

$$\int_a^b f(x)\,\mathrm{d}x \approx h(y_{0.5} + y_{1.5} + \ldots y_{n-0.5}) \text{ where } h = \dfrac{b-a}{n}.$$

Example

Estimate $\displaystyle\int_1^3 \ln x\,\mathrm{d}x$ using the mid-ordinate rule with 4 rectangles.

There are 4 rectangles, so $h = \dfrac{3-1}{4} = 0.5,$

$$x_0 = 1, x_1 = 1.5, x_2 = 2, x_3 = 2.5, x_4 = 3.$$

$$\therefore \quad x_{0.5} = 1.25 \implies y_{0.5} = \ln 1.25.$$

Calculating the other y mid-ordinates in a similar way gives:

$$\int_1^3 \ln x \, dx \approx 0.5(\ln 1.25 + \ln 1.75 + \ln 2.25 + \ln 2.75)$$

$$= 1.3026 \ldots = 1.30 \ (2 \text{ d.p.})$$

Using integration by parts, the exact answer is $3 \ln 3 - 2 = 1.295 \ldots$ so the answers agree to 2 d.p.

Progress check

1 Show that the equation $x^3 - x - 7 = 0$ has a root between 2 and 3. Use a decimal search to find the value of the root to 2 d.p.

2 Use simple iteration to refine the value of the root found in question 1 and state its value to 4 d.p.

3 Use the trapezium rule with five strips to find the value of $\int_2^3 x \ln x \, dx$ to 2 d.p.

4 Show that the equation $e^x + 2x^2 - 2 = 0$ has a root between $x = 0.4$ and $x = 0.5$.

Using the Newton-Raphson method with first approximation 0.5, find the value of the root correct to 2 decimal places.

5 Estimate $\int_0^1 \sqrt{1 - x^3} \, dx$

(a) using Simpson's rule with 4 strips
(b) using the mid-ordinate rule with 4 rectangles
(c) using the trapezium rule with 4 trapezia.

AQA
EDEXCEL

1 2.09
2 2.0867
3 2.31
4 0.46
5 To 3 d.p. (a) 0.823 (b) 0.854 (c) 0.797

59

1.8 Vectors

After studying this section you should be able to:

- *understand the distinction between vector and scalar quantities*
- *add and subtract vector quantities and multiply by a scalar*
- *use the unit vectors **i**, **j** and **k***
- *find the magnitude and direction of a vector*
- *use scalar products to find the angle between two directions*
- *express equations of lines in vector form*
- *determine whether two lines are parallel, intersect or are skew*

LEARNING SUMMARY

Vector and scalar quantities

AQA	C4
EDEXCEL	C4
OCR	C4
WJEC	C4
CCEA	C4

A **scalar** quantity has size (or **magnitude**) but not direction. **Numbers** are scalars and some other important examples are **distance**, **speed**, **mass** and **time**.

A **vector** quantity has both size and **direction**. For example, distance in a specified direction is called **displacement**.

The diagram shows a **directed line segment**. It has size (in this case, length) and direction so it is a vector.

The diagram gives a useful way to represent *any* vector quantity and may also be used to represent addition and subtraction of vectors and multiplication of a vector by a scalar.

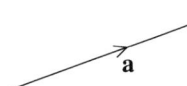

In a textbook, vectors are usually labelled with lower case letters in **bold** print.
When hand-written, these letters should be underlined e.g. $\underline{a}$.

Addition and subtraction of vectors

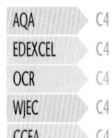

AQA	C4
EDEXCEL	C4
OCR	C4
WJEC	C4
CCEA	C4

This diagram shows three vectors **a**, **b** and **c** such that **c** = **a** + **b**.

The vectors **a** and **b** follow on from each other and then **c** joins the start of vector **a** to the end of vector **b**.

Check that **a** + **b** = **b** + **a**.

The vector **c** is called the **resultant** of **a** and **b**.

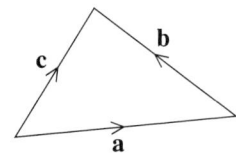

On a vector diagram, −**q** has the opposite sense of direction to **q**.
Notice that **p** − **q** is represented as **p** + (−**q**).

This diagram shows the same information in a different way.

Following the route in the opposite direction to **q** is the same as adding −**q** or subtracting **q**.

Scalar multiplication

AQA | C4
EDEXCEL | C4
OCR | C4
WJEC | C4
CCEA | C4

2**p** is parallel to **p** and has the same sense of direction but is twice as long.

−3**p** is parallel to **p** but has the opposite sense of direction and is three times as long.

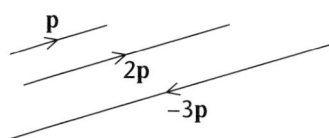

Component form

AQA | C4
EDEXCEL | C4
OCR | C4
WJEC | C4
CCEA | C4

When working in two dimensions, it is often very useful to express a vector in terms of two special vectors **i** and **j**. These are **unit vectors** at right-angles to each other.
A vector **r** written as **r** = a**i** + b**j** is said to have **components** a**i** and b**j**.

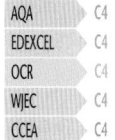

> A unit vector is a vector of magnitude 1 unit.

Examples

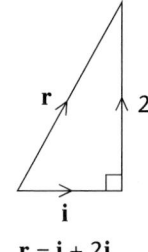

r = **i** + 2**j**

r = 3**i** − **j**

For work involving the Cartesian coordinate system, **i** and **j** are taken to be in the positive directions of the x- and y-axes respectively.

In three dimensions, a third vector **k** is used to represent a unit vector in the positive direction of the z-axis.

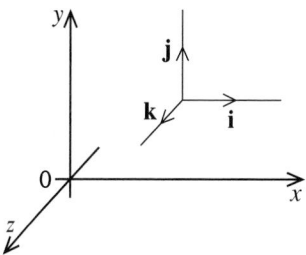

The **position vector** of a point P is the vector $\overrightarrow{OP}$ where O is the origin.

If P has coordinates (a, b) then its position vector is given by **r** = a**i** + b**j**.

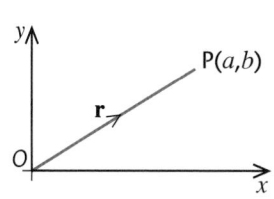

In three dimensions, a point with coordinates (a, b, c) would have position vector **r** = a**i** + b**j** + c**k**.

Adding and subtracting vectors in component form

AQA | C4
EDEXCEL | C4
OCR | C4
WJEC | C4
CCEA | C4

One advantage of expressing vectors in terms of **i**, **j** and **k** is that addition and subtraction may be done algebraically.

Example
p = 3**i** + 2**j** − **k** and **q** = 5**i** − **j** + 4**k**. Express the following vectors in terms of **i**, **j** and **k**:

(a) **p** + **q** (b) **p** − **q** (c) 2**p** − 3**q**.

(a) **p** + **q** = (3**i** + 2**j** − **k**) + (5**i** − **j** + 4**k**) = 8**i** + **j** + 3**k**

(b) **p** − **q** = (3**i** + 2**j** − **k**) − (5**i** − **j** + 4**k**) = − 2**i** + 3**j** − 5**k**

(c) 2**p** − 3**q** = 2(3**i** + 2**j** − **k**) − 3(5**i** − **j** + 4**k**)
   = 6**i** + 4**j** − 2**k** − 15**i** + 3**j** − 12**k** = −9**i** + 7**j** − 14**k**.

Finding the magnitude of a vector

AQA	C4
EDEXCEL	C4
OCR	C4
WJEC	C4
CCEA	C4

> **KEY POINT**
>
> If $\mathbf{r} = a\mathbf{i} + b\mathbf{j} + c\mathbf{k}$, the magnitude (length) of $\mathbf{r}$ is given by
> $$|\mathbf{r}| = \sqrt{a^2 + b^2 + c^2}$$

For example,
If $\mathbf{r} = 3\mathbf{i} - 4\mathbf{j} + 12\mathbf{k}$, then $|\mathbf{r}| = \sqrt{3^2 + (-4)^2 + 12^2} = 13$.

Scalar product

AQA	C4
EDEXCEL	C4
OCR	C4
WJEC	C4
CCEA	C4

If $\mathbf{a} = x_1\mathbf{i} + y_1\mathbf{j} + z_1\mathbf{k}$, then $|\mathbf{a}| = \sqrt{x_1^2 + y_1^2 + z_1^2}$ where $|\mathbf{a}|$ is the length of $\mathbf{a}$.

If $\mathbf{b} = x_2\mathbf{i} + y_2\mathbf{j} + z_2\mathbf{k}$, then $|\mathbf{b}| = \sqrt{x_2^2 + y_2^2 + z_2^2}$.

> **KEY POINT**
>
> If θ is the angle between $\mathbf{a}$ and $\mathbf{b}$,
> the scalar (or dot) product of $\mathbf{a}$ and $\mathbf{b}$ is $\mathbf{a.b}$
> where $\mathbf{a.b} = |\mathbf{a}||\mathbf{b}|\cos\theta$.

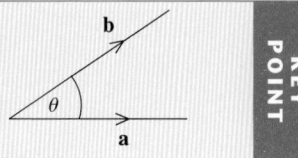

This leads to the useful results:
$\mathbf{i.i} = 1$, $\mathbf{j.j} = 1$, $\mathbf{k.k} = 1$
$\mathbf{i.j} = \mathbf{j.k} = \mathbf{k.i} = 0$.

If two vectors are parallel, $\theta = 0$ and $\mathbf{a.b} = |\mathbf{a}||\mathbf{b}|$.
If two vectors are perpendicular, $\theta = 90°$ and $\mathbf{a.b} = 0$.

For $\mathbf{a}$ and $\mathbf{b}$ defined as above:

In column vectors
$$\mathbf{a.b} = \begin{pmatrix} x_1 \\ y_1 \\ z_1 \end{pmatrix} \cdot \begin{pmatrix} x_2 \\ y_2 \\ z_2 \end{pmatrix}$$
$$= x_1 x_2 + y_1 y_2 + z_1 z_2.$$

$$\mathbf{a.b} = (x_1\mathbf{i} + y_1\mathbf{j} + z_1\mathbf{k}) \cdot (x_2\mathbf{i} + y_2\mathbf{j} + z_2\mathbf{k})$$
$$= x_1 x_2 + y_1 y_2 + z_1 z_2$$

This uses the results for parallel and perpendicular vectors.

Example
Find the angle between $\mathbf{a} = 2\mathbf{i} + \mathbf{j} + 4\mathbf{k}$ and $\mathbf{b} = -3\mathbf{i} + 2\mathbf{j} - \mathbf{k}$.

$$\mathbf{a.b} = 2(-3) + 1(2) + 4(-1) = -8$$
$$|\mathbf{a}| = \sqrt{2^2 + 1^2 + 4^2} = \sqrt{21}$$
$$|\mathbf{b}| = \sqrt{(-3)^2 + 2^2 + (-1)^2} = \sqrt{14}$$
$$\mathbf{a.b} = |\mathbf{a}||\mathbf{b}|\cos\theta$$
$$-8 = \sqrt{21}\,\sqrt{14}\cos\theta$$
$$\Rightarrow \cos\theta = \frac{-8}{\sqrt{21}\,\sqrt{14}} = -0.4665\ldots$$
$$\Rightarrow \theta = 117.8° \ (1 \text{ d.p.})$$

In column vectors:
$$\mathbf{a.b} = \begin{pmatrix} 2 \\ 1 \\ 4 \end{pmatrix} \cdot \begin{pmatrix} -3 \\ 2 \\ -1 \end{pmatrix} = -8.$$

Vector equation of a line

AQA	C4
EDEXCEL	C4
OCR	C4
WJEC	C4
CCEA	C4

> **KEY POINT**
>
> A vector equation of a line passing through a fixed point A with position vector $\mathbf{a}$ and parallel to a vector $\mathbf{b}$ is
> $\mathbf{r} = \mathbf{a} + t\mathbf{b}$, where t is a scalar parameter.

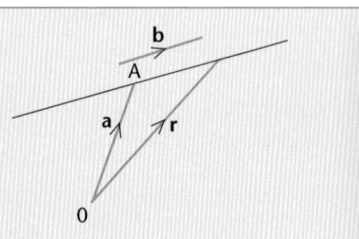

t is often used for the parameter, but not exclusively so. Other letters sometimes used include s, μ and λ.

r is the position vector of any point on the line.
b is the direction vector of the line.

The equation of the line is not unique. Any other point on the line could be used instead of A and any multiple of **b** could be used for the direction vector.

In column format:
$$\mathbf{r} = \begin{pmatrix} 4 \\ -5 \\ 1 \end{pmatrix} + t \begin{pmatrix} 3 \\ 4 \\ -2 \end{pmatrix}.$$

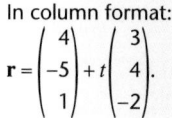

Example
A vector equation of the line passing through $(4, -5, 1)$, parallel to $3\mathbf{i} + 4\mathbf{j} - 2\mathbf{k}$ is $\mathbf{r} = 4\mathbf{i} - 5\mathbf{j} + \mathbf{k} + t(3\mathbf{i} + 4\mathbf{j} - 2\mathbf{k})$.

> **KEY POINT**
> A vector equation of the line through two fixed points A and B, with position vectors **a** and **b** is given by $\mathbf{r} = \mathbf{a} + t(\mathbf{b} - \mathbf{a})$ where t is a scalar parameter.

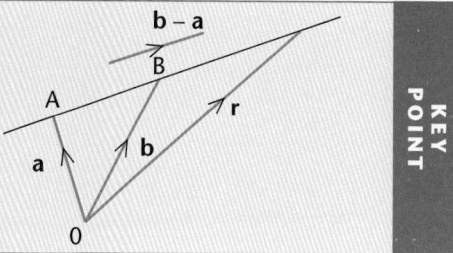

Example
Find a vector equation of the line through the points $P(3, 1, -4)$ and $Q(-2, 5, 1)$.
Direction of line PQ:

$$\overrightarrow{PQ} = \mathbf{q} - \mathbf{p} = -2\mathbf{i} + 5\mathbf{j} + \mathbf{k} - (3\mathbf{i} + \mathbf{j} - 4\mathbf{k}) = -5\mathbf{i} + 4\mathbf{j} + 5\mathbf{k}$$

A vector equation of PQ is $\mathbf{r} = 3\mathbf{i} + \mathbf{j} - 4\mathbf{k} + t(-5\mathbf{i} + 4\mathbf{j} + 5\mathbf{k})$.

Angles between two lines

To find the angle between two lines, find the angle between their direction vectors.

Pairs of lines

AQA   C4
EDEXCEL   C4
OCR   C4
CCEA   C4

In 2-dimensions, a pair of lines are either parallel or they intersect.
In 3-dimensions, a pair of lines are parallel, or they intersect or they are skew.

If two lines are parallel, then their direction vectors are multiples of each other, as in this **example**:
$\mathbf{r} = 3\mathbf{i} - 5\mathbf{j} + 2\mathbf{k} + \lambda(\mathbf{i} + 2\mathbf{j} - \mathbf{k})$ and $\mathbf{r} = 4\mathbf{i} + \mathbf{j} - 2\mathbf{k} + \mu(3\mathbf{i} + 6\mathbf{j} - 3\mathbf{k})$.

The two lines $\mathbf{r} = \mathbf{a} + \lambda\mathbf{b}$ and $\mathbf{r} = \mathbf{c} + \mu\mathbf{d}$ intersect if unique values of λ and μ can be found such that $\mathbf{a} + \lambda\mathbf{b} = \mathbf{c} + \mu\mathbf{d}$.

CCEA does not need skew lines.

If unique values cannot be found, then the two lines are **skew**.

Example
Investigate whether the lines $\mathbf{r}_1 = 2\mathbf{i} + \mathbf{j} - 3\mathbf{k} + \lambda(4\mathbf{i} + 6\mathbf{j} - \mathbf{k})$ and $\mathbf{r}_2 = 3\mathbf{i} - 2\mathbf{k} + \mu(4\mathbf{i} + \mathbf{j} + 3\mathbf{k})$ intersect or whether they are skew.

If the lines intersect, then $\mathbf{r}_1 = \mathbf{r}_2$ will have a unique solution for λ and μ.

$\mathbf{r}_1 = \mathbf{r}_2 \Rightarrow 2\mathbf{i} + \mathbf{j} - 3\mathbf{k} + \lambda(4\mathbf{i} + 6\mathbf{j} - \mathbf{k}) = 3\mathbf{i} - 2\mathbf{k} + \mu(4\mathbf{i} + \mathbf{j} + 3\mathbf{k})$

i.e. $(2 + 4\lambda)\mathbf{i} + (1 + 6\lambda)\mathbf{j} + (-3 - \lambda)\mathbf{k} = (3 + 4\mu)\mathbf{i} + (0 + \mu)\mathbf{j} + (-2 + 3\mu)\mathbf{k}$

Note that the lines are not parallel, since their direction vectors are not multiples of each other.

Equating coefficients:

		Solving (1) and (2) simultaneously gives:
$2 + 4\lambda = 3 + 4\mu$	(1)	$\lambda = -0.25$ and $\mu = -0.5$
$1 + 6\lambda = \mu$	(2)	Check whether these satisfy (3):
$-3 - \lambda = -2 + 3\mu$	(3)	LHS $= -3 - (-0.25) = -2.75$
		RHS $= -2 + 3(-0.5) = -3.5$

$\therefore \lambda = -0.25$ and $\mu = -0.5$ do not satisfy all three equations.

Since there is not a unique solution for λ and μ, the lines do not intersect. They are skew.

Progress check

1 Write down the position vector **r** of a point with coordinates (5, −2).

2 Find the magnitude of the vector **r** = 5**i** − 12**j** and give its direction relative to **i**.

3 Find the angle between the vectors **a** = 2**i** + 3**j** − **k** and **b** = **i** + **j** + 2**k**.

4 (a) Give a vector equation of the line which is parallel to 4**i** − **j** + 2**k** and goes through the point with coordinates (3, 0, 1).

   (b) Give a vector equation of the line through (4, −1, 1) and (−3, 2, −2).

5 (a) Find the point of intersection of the lines

   r = 2**i** − 3**j** + μ(**i** − 2**j**) and **r** = **i** − 2**j** + λ(2**i** + 3**j**).

   (b) Investigate whether the following lines intersect or whether they are skew. If they intersect, find their point of intersection.

   r = **i** + 2**j** − 5**k** + s(3**i** + **j** + **k**) and **r** = 2**i** − **j** + 2**k** + t(−**i** − **j** + **k**)

   (c) Find the angle between the lines given in (b).

5 (a) $(1\tfrac{1}{2}, -1\tfrac{3}{4})$ (b) Intersect at (7, 4, −3) (c) 121.5°
4 (a) **r** = 3**i** + **k** + t(4**i** − **j** + 2**k**) (b) **r** = 4**i** − **j** + **k** + t(−7**i** + 3**j** − 3**k**)
   (Answers are not unique, other answers are possible.)
3 70.9°
2 |**r**| = 13, 67.4° clockwise.
1 **r** = 5**i** − 2**j**

Sample questions and model answers

1

(a) Express $f(x) = \dfrac{5x+1}{(x+2)(x-1)^2}$ in partial fractions.

(b) Hence show that $\displaystyle\int_{-1}^{0} f(x)\,dx = 1 - 2\ln 2$.

Remember the format when there is a repeated factor.

Take care with the denominator.

(a) Let $\dfrac{5x+1}{(x+2)(x-1)^2} \equiv \dfrac{A}{x+2} + \dfrac{B}{x-1} + \dfrac{C}{(x-1)^2}$

$\Rightarrow \dfrac{5x+1}{(x+2)(x-1)^2} \equiv \dfrac{A(x-1)^2 + B(x+2)(x-1) + C(x+2)}{(x+2)(x-1)^2}$

$\Rightarrow 5x+1 \equiv A(x-1)^2 + B(x+2)(x-1) + C(x+2)$

Putting $x = 1$ makes the factor $(x-1)$ equal to zero.

Let $x = 1$: $\quad 6 = 3C \Rightarrow C = 2$

You could equate the x term or the constant term if you wish.

Let $x = -2$: $\quad -9 = 9A \Rightarrow A = -1$

Equate x^2 terms: $\quad 0 = A + B \Rightarrow B = 1$

$\therefore f(x) = -\dfrac{1}{x+2} + \dfrac{1}{x-1} + \dfrac{2}{(x-1)^2}$

(b) $\displaystyle\int_{-1}^{0} f(x)\,dx = \int_{-1}^{0}\left(-\dfrac{1}{x+2} + \dfrac{1}{x-1} + \dfrac{2}{(x-1)^2}\right)dx$

Remember the modulus sign.

$= \left[-\ln|x+2| + \ln|x-1| - 2(x-1)^{-1}\right]_{-1}^{0}$

$= -\ln 2 + \ln|-1| + 2 - (-\ln 1 + \ln|-2| + 1)$

$= -\ln 2 + 2 - \ln 2 - 1$

$\ln 1 = \ln|-1| = 0$
$\ln|-2| = \ln 2$.

$= 1 - 2\ln 2$

2

Expand $(1 - 2x)^{-3}$ in ascending powers of x as far as the term in x^3 and state the set of values of x for which the expansion is valid.

Take care with the negatives.

Remember to include the restriction on x, even when it is not requested specifically in the question.

$(1 - 2x)^{-3} = 1 + (-3)(-2x) + \dfrac{(-3)(-4)}{2!}(-2x)^2 + \dfrac{(-3)(-4)(-5)}{3!}(-2x)^3 + \dots$

$= 1 + 6x + 24x^2 + 80x^3 + \dots$ provided $|2x| < 1$, i.e. $|x| < 0.5$

Sample questions and model answers (continued)

3

Use the substitution $x = \tan u$ to show that $\displaystyle\int_0^1 \frac{1}{(1+x^2)^2}\,dx = \int_0^{\frac{\pi}{4}} \cos^2 u\,du$.

Hence show that $\displaystyle\int_0^1 \frac{1}{(1+x^2)^2}\,dx = \frac{\pi}{8} + \frac{1}{4}$.

> Remember to change the x limits to u limits.

> To integrate even powers of $\sin x$ or $\cos x$, change to double angles using $\cos 2x$ formula.

$$\int_0^1 \frac{1}{(1+x^2)^2}\,dx = \int_{x=0}^{x=1} \frac{1}{(1+x^2)^2} \frac{dx}{du}\,du$$

$$= \int_0^{\frac{\pi}{4}} \frac{1}{(\sec^2 u)^2} \sec^2 u\,du$$

$$= \int_0^{\frac{\pi}{4}} \frac{1}{\sec^2 u}\,du$$

$$= \int_0^{\frac{\pi}{4}} \cos^2 u\,du$$

> Show this essential side working.

Let $x = \tan u$

$\Rightarrow \dfrac{dx}{du} = \sec^2 u$

$1 + x^2 = 1 + \tan^2 u$
$\qquad\ = \sec^2 u$

Limits:

When $x = 0$, $\tan u = 0 \Rightarrow u = 0$

When $x = 1$, $\tan u = 1 \Rightarrow u = \dfrac{\pi}{4}$

$$\int_0^{\frac{\pi}{4}} \cos^2 u\,du = \frac{1}{2}\int_0^{\frac{\pi}{4}} (1 + \cos 2u)\,du$$

$$= \frac{1}{2}\left[u + \frac{1}{2}\sin 2u \right]_0^{\frac{\pi}{4}}$$

$$= \frac{1}{2}\left(\frac{\pi}{4} + \frac{1}{2} - 0 \right)$$

$$= \frac{\pi}{8} + \frac{1}{4}$$

$\cos 2u = 2\cos^2 u - 1$

$\Rightarrow \cos^2 u = \frac{1}{2}(1 + \cos 2u)$

4

(a) Solve the inequality $|2x - 3| < 11$

(b) Sketch the graph of $y = |f(x)|$ where $f(x) = (x - 1)(x - 2)(x - 3)$

(a) $|2x - 3| < 11$

> This first step is important.

$\Rightarrow -11 < 2x - 3 < 11$

> You can simplify the inequality by doing the same thing to each part.

$\Rightarrow -8 < 2x < 14$

$\Rightarrow -4 < x < 7$.

> Add 3 to each part.

> Divide each part by 2.

(b)

> Parts of the graph of $y = f(x)$ that would be below the x-axis are reflected in the x-axis to give the graph of $y = |f(x)|$.

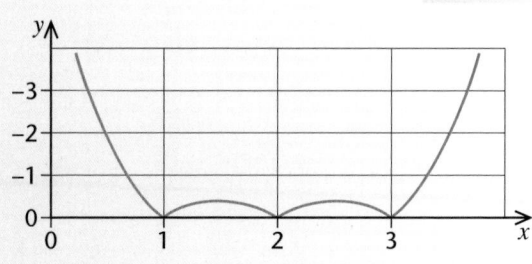

Sample questions and model answers (continued)

5

The functions f and g are defined by:

$$f(x) = \sqrt{x-1} \quad x \geqslant 1$$
$$g(x) = 2x + 3 \quad x \in \mathbb{R}$$

(a) Find an expression for $f^{-1}(x)$ and state its domain and range.

(b) Find the value of:

   (i) $fg(7)$

   (ii) $gf(10)$

   (iii) $f^{-1}g^{-1}(15)$

Set $y = f(x)$ and rearrange to make x the subject.

(a) $y = \sqrt{x-1}$

$\Rightarrow y^2 = x - 1$

$\Rightarrow x = y^2 + 1$

$\Rightarrow f^{-1}(x) = x^2 + 1$

The domain of f^{-1} is given by the range of f.

The domain of f^{-1} is $x \geqslant 0$ and the range of f^{-1} is $x \geqslant 1$.

The range of f^{-1} is given by the domain of f.

The order in which you apply the functions is important.

(b) (i) $fg(7) = f(17) = 4$

(ii) $gf(10) = g(3) = 9$

(iii) $f^{-1}g^{-1}(15) = f^{-1}(6) = 37$

6

The equation of a straight line l is

$$\mathbf{r} = 4\mathbf{i} - 2\mathbf{k} + t(3\mathbf{i} + \mathbf{k}) \qquad \text{where } t \text{ is a scalar parameter.}$$

Points A, B and C lie on the line l. Referred to the origin O, they have position vectors $\mathbf{a}$, $\mathbf{b}$ and $\mathbf{c}$ respectively.

(a)   Find $\mathbf{a}$ and $\mathbf{b}$ and angle AOB, given that distance $OA = OB = 10$ units.

(b)   Find $\mathbf{c}$, given that OC is perpendicular to l.

(a) $\underline{a} = (4 + 3t)\underline{i} + (-2 + t)\underline{k}$ for some value t.   (1)

Distance $OA = |\mathbf{a}|$.

Since $|\underline{a}| = 10$,

$$\sqrt{(4 + 3t)^2 + (-2 + t)^2} = 10$$

$$16 + 24t + 9t^2 + 4 - 4t + t^2 = 100$$

$$10t^2 + 20t - 80 = 0$$

$$t^2 + 2t - 8 = 0$$

$$(t - 2)(t + 4) = 0$$

$$\Rightarrow t = 2 \text{ or } t = -4$$

Sample questions and model answers (continued)

This gives the position vectors of **a** and **b**.

Substituting for t in (1) gives

$$\underline{a} = 10\underline{i} \quad \text{and} \quad \underline{b} = -8\underline{i} - 6\underline{k}$$

Let angle AOB be θ.

$$|\underline{a}| = |\underline{b}| = 10 \quad \text{and} \quad \underline{a} \cdot \underline{b} = (10\underline{i}) \cdot (-8\underline{i} - 6\underline{k}) = -80$$

$$\underline{a} \cdot \underline{b} = |\underline{a}| \, |\underline{b}| \cos \theta$$

$$\Rightarrow \quad \cos \theta = \frac{-80}{100} = -0.8$$

$$\Rightarrow \quad \theta = 143.1° \text{ (1 d.p.)}$$

(b) $\quad \underline{c} = (4 + 3t)\underline{i} + (-2 + t)\underline{k} \quad$ for some value t. $\qquad$ (2)

3i + **k** is the direction vector of l and the scalar product of perpendicular vectors is zero.

Since $\underline{c}$ is perpendicular to l,

$$((4 + 3t)\underline{i} + (-2 + t)\underline{k}) \cdot (3\underline{i} + \underline{k}) = 0$$

$$\Rightarrow \quad 12 + 9t - 2 + t = 0$$

$$\Rightarrow \quad t = -1$$

Substituting for t in (2) gives $\underline{c} = \underline{i} - 3\underline{k}$.

Practice examination questions

1 It is given that $f(x) = \dfrac{1}{(1+x)^2} + \sqrt{4+x}$.

Show that if x^3 and higher powers are ignored, $f(x) \approx a + bx + cx^2$ and find the values of a, b and c.

State the values of x for which the expansion is valid.

2 (a) Find $\dfrac{dy}{dx}$ when $y = \cos^2 x \sin x$, expressing your answer in terms of $\cos x$.

(b) Using the substitution $u = \cos x$, or otherwise, find $\displaystyle\int \cos^2 x \sin x \, dx$.

3 A curve has equation $y = \dfrac{x^2 - 4}{x + 1}$.

The equation of the normal to the curve at $(2, 0)$ is $ax + by + c = 0$, where a, b and c are integers.

Find the values of a, b, and c.

4 The functions f and g are defined by:

$$f{:}x \mapsto 5x - 7 \qquad x \in \mathbb{R}$$

$$g{:}x \mapsto (x+1)(x-1) \quad x \in \mathbb{R}.$$

(a) Find the range of g.

(b) Find an expression for $fg(x)$.

(c) Determine the values of x for which $fg(x) = f(x)$.

5 (a) Express $15 \sin x + 8 \cos x$ in the form $R \sin(x + a)$ where $0 < a < 90°$.

(b) State the maximum value of $15 \sin x + 8 \cos x$.

(c) Solve the equation $15 \sin x + 8 \cos x = 12$.

Give all the solutions in the interval $0 < x < 360°$.

6 Expressing y in terms of x, solve the differential equation

$$x \frac{dy}{dx} = y + yx,$$

given that $y = 4$ when $x = 2$.

7 (a) Use the chain rule to differentiate:

(i) $y = (4x - 3)^9$

(ii) $y = \ln(5 - 2x)$

(b) Using your results from part (a):

(i) Write down an expression for $\displaystyle\int (4x - 3)^8 \, dx$

(ii) Calculate $\displaystyle\int_3^4 \frac{1}{5 - 2x} \, dx$

Practice examination questions (continued)

8 (a) Use integration by parts to evaluate $\int_0^1 x\,e^{-2x}\,dx$.

(b) The diagram shows the graph of $y = x\,e^{-x}$.

The curve has a maximum point at A.

The shaded region R is the area bounded by the curve, the x-axis and the vertical line through A.

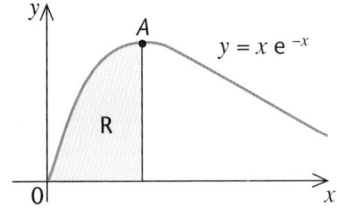

(i) Show that the coordinates of A are $(1, e^{-1})$.
(ii) The region R is rotated completely about the x-axis.
Calculate the volume of the solid generated.

9 In a biology experiment, the growth of a population is being investigated.

A simple model is set up in which the rate of increase of the population at time t is proportional
to the number, P, in the population at that time.

It is known that when $t = 0$, $P = 500$ and when $t = 10$, $P = 1000$.

Obtain the relationship between P and t and estimate the size of the population, according to the model, when $t = 20$.

10 (a) Find the equation of the normal to the curve $x^3 + xy + y^2 = 7$ at the point $(1, 2)$.

(b) Find the coordinates of the two stationary points on the curve $x = t^2$, $y = t + \dfrac{1}{t}$.

11 The region R, shaded in the diagram, is bounded by the curve $y = \sin 2x$ and the x-axis.

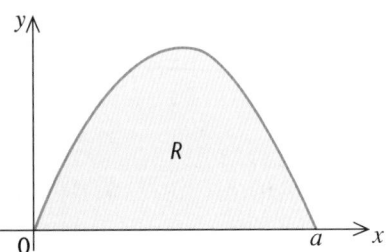

(a) Find the value of a.

(b) Show that the area of R is 1 square unit.

(c) Find the volume of the solid formed when R is rotated completely about the x-axis.

12 (a) Find a vector equation of the line AB through $A(2, 2, 0)$ and $B(2, 3, 1)$.

(b) Show that the line AB and the line $\mathbf{r} = 2\mathbf{j} + \mathbf{k} + \lambda(\mathbf{i} + \mathbf{k})$, where λ is a scalar parameter, have no common point.

(c) Find the acute angle between the two lines.

Chapter 2

Statistics 2

The following topics are covered in this chapter:

- Continuous random variables
- Statistical approximations
- Estimation and sampling
- Hypothesis tests 1
- Hypothesis tests 2

2.1 Continuous random variables

After studying this section you should be able to:

- find probabilities and the mean and variance of a continuous variable with probability density function $f(x)$
- find the cumulative function $F(x)$ and the median and quartiles
- find the probability density function from the cumulative distribution function
- use the rules relating to linear combinations of variables, especially normal variables
- understand the continuous uniform distribution

LEARNING SUMMARY

The probability density function, $f(x)$

AQA	S2
EDEXCEL	S2
OCR	S2
WJEC	S1
CCEA	S1

The continuous random variable X is defined by its **probability density function** $f(x)$, together with the values for which it is valid.

$$P(c \leqslant x \leqslant d) = \int_c^d f(x)\,dx$$

The total probability is 1.

If $f(x)$ is valid for $a \leqslant x \leqslant b$, then $\int_a^b f(x)\,dx = 1$.

Probabilities are given by areas under the curve.

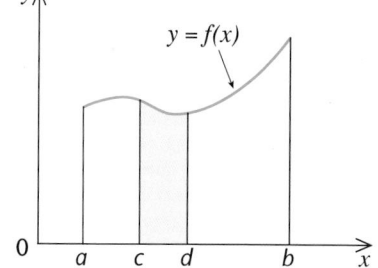

The **expectation** of $g(x)$ is $E(g(x)) = \int_{\text{all } x} g(x)f(x)\,dx$, for any function $g(x)$.

The functions used most often are X and X^2 as $E(X)$ is the mean and $E(X^2)$ is used to find the variance.

The mean is the **expectation** of X, written $E(X)$. It is often denoted by μ.

If $f(x)$ is given in stages, then the integration is carried out in stages also.

KEY POINT

Mean: $\qquad E(X) = \mu = \int_{\text{all } x} xf(x)\,dx$

Variance: $\quad \text{Var}(X) = \sigma^2 = E(X^2) - \mu^2$, where $E(X^2) = \int_{\text{all } x} x^2 f(x)\,dx$

Example

The continuous random variable X has probability density function $f(x)$ given by $f(x) = \frac{1}{4}x$ for $1 \leqslant x \leqslant 3$. Find the mean μ, $E(X^2)$ and the standard deviation σ.

$$\mu = E(X) = \int_{\text{all } x} xf(x)\,dx = \int_1^3 \frac{1}{4}x^2\,dx = \left[\frac{x^3}{12}\right]_1^3 = 2\tfrac{1}{6}$$

$$E(X^2) = \int_{\text{all } x} x^2 f(x)\,dx = \int_1^3 \frac{1}{4}x^3\,dx = \left[\frac{x^4}{16}\right]_1^3 = 5$$

$\sigma = \sqrt{\text{variance}}.$

$$\text{Var}(X) = 5 - (2\tfrac{1}{6})^2 = \tfrac{11}{36} \implies \sigma = \sqrt{\tfrac{11}{36}} = 0.553 \text{ (3 d.p.)}$$

The cumulative distribution function, $F(x)$

AQA   S2
EDEXCEL   S2
OCR   S2
WJEC   S1

The **cumulative distribution function**, $F(x)$, gives the probability that X is less than a particular value, i.e. $P(X < x)$. It is found from the area under the probability curve:

$F(x_1) = P(X < x_1)$.

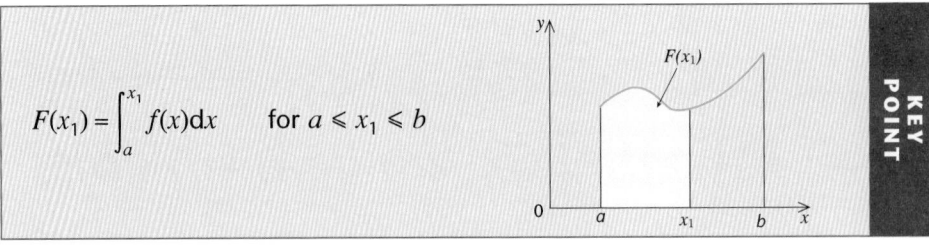

$$F(x_1) = \int_a^{x_1} f(x)\,dx \qquad \text{for } a \leqslant x_1 \leqslant b$$

KEY POINT

Sometimes the median is denoted by q_2.

The **median**, m, is the value 50% of the way through the distribution. It divides the total area in half, so $F(m) = 0.5$.

The **lower quartile** q_1 is the value 25% of the way, so $F(q_1) = 0.25$.
The **upper quartile** q_3 is the value 75% of the way, so $F(q_3) = 0.75$.

The **interquartile range** = upper quartile − lower quartile.

Example

$f(x) = \frac{1}{4}x$ for $1 \leqslant x \leqslant 3$. Find the median, m, and the interquartile range.

x_1 is used here to denote any value of x between 1 and 3. Sometimes a different letter is used, such as t.

For $1 \leqslant x_1 \leqslant 3$, $F(x_1) = \int_1^x \frac{1}{4}x\,dx = \left[\dfrac{x^2}{8}\right]_1^{x_1} = \frac{1}{8}(x_1^2 - 1)$

$F(m) = 0.5 \Rightarrow \frac{1}{8}(m^2 - 1) = 0.5 \Rightarrow m^2 = 5,\ m = \sqrt{5}\ (= 2.236\ldots)$
$F(q_1) = 0.25 \Rightarrow \frac{1}{8}(q_1^2 - 1) = 0.25 \Rightarrow q_1^2 = 3,\ q_1 = \sqrt{3}\ (= 1.732\ldots)$
$F(q_3) = 0.75 \Rightarrow \frac{1}{8}(q_3^2 - 1) = 0.75 \Rightarrow q_3^2 = 7,\ q_3 = \sqrt{7}\ (= 2.645\ldots)$
Interquartile range $= q_3 - q_1 = \sqrt{7} - \sqrt{3} = 0.91$ (2 d.p.)

Differentiating the cumulative function gives the probability function.

The probability density function can be obtained from the cumulative distribution function:

$$f(x) = \frac{d}{dx}F(x).$$

KEY POINT

Example

Find $f(x)$ where $F(x) = \begin{cases} 0 & x \leqslant 0 \\ \frac{1}{8}x^3 & 0 \leqslant x \leqslant 2 \\ 1 & x \geqslant 2 \end{cases}$

$f(x) = \dfrac{d}{dx}(\frac{1}{8}x^3) = \frac{3}{8}x^2$ for $0 \leqslant x \leqslant 2$ and $f(x) = 0$ otherwise.

Linear combinations of random variables

AQA   S2
EDEXCEL   S1
WJEC   S2
CCEA   S2/S1

The following results relating to **expectation algebra**, hold for continuous and discrete variables.

In these examples, $b = -4$.

For variable X and constants a and b:

$E(aX) = aE(X)$
$E(aX + b) = aE(X) + b$

$\text{Var}(aX) = a^2\,\text{Var}(X)$
$\text{Var}(aX + b) = a^2\,\text{Var}(X)$

Examples

$E(3X) = 3E(X)$
$E(5X - 4) = 5E(X) - 4$

$\text{Var}(3X) = 9\,\text{Var}(X)$
$\text{Var}(5X - 4) = 25\,\text{Var}(X)$

For variables X and Y:

$E(aX + bY) = aE(X) + bE(Y)$

$E(aX - bY) = aE(X) - bE(Y)$

If X and Y are independent:

Remember the + sign in the variance result for $\text{Var}(aX - bY)$.

$\text{Var}(aX + bY) = a^2\,\text{Var}(X) + b^2\,\text{Var}(Y)$

$\text{Var}(aX - bY) = a^2\,\text{Var}(X) + b^2\,\text{Var}(Y)$

Examples

$E(2X + 3Y) = 2E(X) + 3E(Y)$

$E(2X - 3Y) = 2E(X) - 3E(Y)$

$\text{Var}(2X + 3Y) = 4\,\text{Var}(X) + 9\,\text{Var}(Y)$

$\text{Var}(2X - 3Y) = 4\,\text{Var}(X) + 9\,\text{Var}(Y)$

Learn this very important result.

If X and Y are **normally distributed**, then sums, differences and multiples of these normal variables are **also normally distributed**.

In general, if $X \sim N(\mu_1, \sigma_1^2)$, $Y \sim N(\mu_2, \sigma_2^2)$ then

$$X + Y \sim N(\mu_1 + \mu_2, \sigma_1^2 + \sigma_2^2) \qquad \text{(sum)}$$

$$X - Y \sim N(\mu_1 - \mu_2, \sigma_1^2 + \sigma_2^2) \qquad \text{(difference)}$$

$$aX \sim N(a\mu_1, a^2\sigma_1^2) \qquad \text{(multiple)}$$

Example

$X \sim N(10, 4)$ and $Y \sim N(8, 3)$. Find the distribution of $2X - 3Y$.

$E(2X - 3Y) = 2E(X) - 3E(Y) = 20 - 24 = -4$

$\text{Var}(2X - 3Y) = 4\,\text{Var}(X) + 9\,\text{Var}(Y) = 16 + 27 = 43$

So $2X - 3Y \sim N(-4, 43)$

The uniform distribution

AQA — S2
EDEXCEL — S2
WJEC — S2

If $f(x)$ is distributed **uniformly** in the interval $a \le x \le b$, then $f(x) = \dfrac{1}{b - a}$.

KEY POINT

By symmetry, the mean is mid-way between a and b, i.e. $E(X) = \frac{1}{2}(a + b)$.
It can be shown that $\text{Var}(X) = \frac{1}{12}(b - a)^2$.

Progress check

1. $f(x) = 1 - 0.5x$, $0 \le x \le 2$.
   (a) Find $E(X)$, $E(X^2)$ and $\text{Var}(X)$.
   (b) Find the cumulative function $F(x)$, the median and the interquartile range.

2. $f(x)$ is distributed uniformly in the interval $1 \le x \le 6$. Find:
   (a) $P(1.8 < X < 3.2)$ (b) $E(X)$ (c) the standard deviation of X.

3. $X \sim N(30, 4)$ and $Y \sim N(20, 5)$. Find:
   (a) $P(X + Y > 56)$ (b) $P(X - Y < 7)$ (c) $P(3X - 4Y > -10)$

4. The masses of apples from a particular tree are normally distributed with mean 135 g and standard deviation 25 g. Find the probability that 7 apples from the tree weigh more than 1 kg.

4. 0.163

3. (a) 0.0228 (b) 0.1587 (c) 0.9683

2. (a) 0.28 (b) 3.5 (c) 1.44 (2 d.p.)

IQR = 0.732 (3 d.p.)

1. (a) $\frac{2}{3}$, $\frac{1}{2}$, $\frac{1}{18}$ (b) $F(x) = 0$, $x \le 0$, $F(x) = x - 0.25x^2$, $0 \le x \le 2$, $F(x) = 1$, $x > 2$; median = 0.586 (3 d.p.)

2.2 Statistical approximations

After studying this section you should be able to:

- use the Poisson approximation to the binomial distribution
- use the normal approximation to the binomial distribution
- use the normal approximation to the Poisson distribution

LEARNING SUMMARY

The Poisson approximation to the binomial distribution

EDEXCEL	S2
OCR	S2
WJEC	S1

See Revise AS for more on the binomial and Poisson distributions.

The binomial distribution $X \sim B(n, p)$ is used to model the number of successes in n independent trials when the probability of success, p, is constant.

The mean of the binomial distribution is np and the variance is npq (where $q = 1 - p$).

With these conditions, $np \approx npq$. This fits in with the Poisson distribution where the mean and the variance are equal.

> **KEY POINT**
>
> When n is large ($n > 50$, say) and p is small ($p < 0.1$, say), $X \sim B(n, p)$ can be **approximated** by a Poisson distribution with the same mean, where $X \sim Po(np)$ approximately.

Ideally, np should be less than 5.

Key points from AS

- **Binomial distribution**
  Revise AS page 95
- **Poisson distribution**
  Revise AS page 96

$$P(X = x) = \frac{\lambda^x}{x!} e^{-\lambda}$$

It is often quicker to use cumulative probability tables, so make sure that you know how to use them.

Example

The proportion of defective items produced by a machine is 2.5%. Find the probability of obtaining fewer than 3 defective items in a random sample of 100 items.

If X is the number of defective items in 100, then $X \sim B(100, 0.025)$.
The mean $np = 100 \times 0.025 = 2.5$.
Since n is large and p is small, use the Poisson approximation, $X \sim Po(2.5)$.

$$P(X < 3) = P(X = 0) + P(X = 1) + P(X = 2)$$

$$= e^{-2.5} + 2.5e^{-2.5} + \frac{2.5^2}{2!} e^{-2.5} = e^{-2.5}\left(1 + 2.5 + \frac{2.5^2}{2!}\right) = 0.54 \ (2 \text{ s.f.})$$

If you have access to cumulative Poisson tables giving $P(X \leqslant r)$ for various values of r, then these tables can be used with $\lambda = 2.5$ and
$P(X < 3) = P(X \leqslant 2) = 0.5438 = 0.54 \ (2 \text{ s.f.})$.

The normal approximation to the binomial distribution

EDEXCEL	S2
OCR	S2
WJEC	S2

> **KEY POINT**
>
> When n and p are such that $np > 5$ and $nq > 5$, $X \sim B(n, p)$ can be **approximated** by a normal distribution with the same mean and the same variance, where $X \sim N(np, npq)$ approximately.

Key points from AS

- **The normal distribution**
  Revise AS pages 100

The larger the value of n and the closer that p is to 0.5, the better the approximation.

The normal distribution is continuous, whereas the binomial is discrete, so a **continuity correction** must be used.

Example

The probability that a person supports a particular organisation is 0.3. In a random sample of 50 people, find the probability that at least 20 support the organisation.

Always define the variable, giving its distribution if known.

If X is the number of people in 50 who support the organisation, then $X \sim \mathrm{B}(50, 0.3)$.

Mean $np = 50 \times 0.3 = 15$, variance $npq = 50 \times 0.3 \times 0.7 = 10.5$.

Conditions for normal approximation: $np = 15 > 5$, $nq = 50 \times 0.7 = 35 > 5$

$X \sim \mathrm{N}(\mu, \sigma^2)$ with $\mu = 15$ and $\sigma^2 = 10.5 \Rightarrow \sigma = \sqrt{10.5}$.

$$\therefore \quad X \sim \mathrm{N}(15, 10.5) \text{ approximately.}$$

Applying the continuity correction, $P(X \geqslant 20)$ becomes $P(X > 19.5)$.

Think of $X = 20$ as going from 19.5 to 20.5. Since 'at least 20' includes $X = 20$, find $P(X > 19.5)$ in the normal distribution.

$$P(X > 19.5) = P\left(Z > \frac{19.5 - 15}{\sqrt{10.5}}\right)$$

$$= P(Z > 1.389)$$

$$= 1 - \Phi(1.389) = 1 - 0.9176 = 0.082 \ (2 \text{ s.f.})$$

Z: 0   1.389

Standardise
$$Z = \frac{X - \mu}{\sigma}$$
and then use standard normal tables.

The normal approximation to the Poisson distribution

EDEXCEL   S2
OCR   S2
WJEC   S2,

The Poisson distribution, $X \sim \mathrm{Po}(\lambda)$, is used to model the number of occurrences of an event, when events occur randomly. The mean is λ and the variance is λ.

> **KEY POINT**
> When λ is large ($\lambda > 15$ say), $X \sim \mathrm{Po}(\lambda)$ can be approximated by a normal distribution with the same mean and variance, where $X \sim \mathrm{N}(\lambda, \lambda)$ approximately.

Since the Poisson distribution is discrete, a continuity correction must be used.

Example

If $X \sim \mathrm{Po}(20)$, find $P(15 < X < 22)$.

Since n is large, $X \sim \mathrm{N}(20, 20)$ approximately.

You do not want to include 15 or 22 so go from 15.5 to 21.5.

$P(15 < X < 22)$ becomes $P(15.5 < X < 21.5)$.

Apply the continuity correction.

Standardise the variables and then use normal tables.

$$P(15.5 < X < 21.5) = P\left(\frac{15.5 - 20}{\sqrt{20}} < Z < \frac{21.5 - 20}{\sqrt{20}}\right)$$

$$= P(-1.006 < Z < 0.335)$$

$$= \Phi(1.006) + \Phi(0.335) - 1$$

$$= 0.8427 + 0.6312 - 1$$

$$= 0.474 \ (3 \text{ s.f.})$$

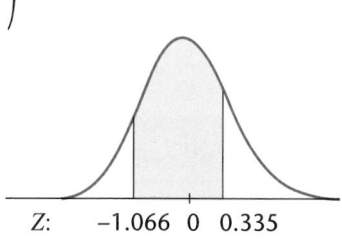

Z: −1.066   0   0.335

Progress check

1. $X \sim \mathrm{B}(80, 0.06)$. Find, to 3 decimal places, $P(X = 4)$
   (a) by using the binomial distribution,
   (b) by using a suitable approximation.

2. A newspaper reports that 62% of adults have an e-mail address. A random sample of 40 adults was selected. Using a suitable approximation, find the probability that at least 30 had an e-mail address.

3. Telephone calls reach a switchboard independently and at random at a rate of 30 per hour. Find the probability that in a randomly selected period of one hour there are fewer than 20 calls.

1 (a) 0.186 (b) 0.182   2 0.063 (2 s.f.)   3 0.028 (2 s.f.)

2.3 Estimation and sampling

Sampling techniques

EDEXCEL	S2
OCR	S2
WJEC	S2
CCEA	S2

When you want information about a **population**, you could carry out a **census** of every member. The advantage is that you would have accurate and complete information. Disadvantages include cost and time requirements. Taking a census could also destroy the population. For example, if you were investigating the length of life of a calculator battery, carrying out a census would destroy the population.

More usually, a **sample** is taken. The sample should be representative of the whole population, so **bias** in the choice of sample members must be avoided.

In a **random sample** of size n, each member of the population has an equal chance of being selected. Also all subsets of the population of size n have an equal chance of constituting the sample.

Assigning a number to each member of a population and then drawing the numbers out of a hat is one method of obtaining a **simple random sample**. Another is to use **random number tables**, in which each digit has an equal chance of occurring.

If there is a periodic fault, then systematic sampling may fail to register it or may give it undue weight.

Random sampling from a very large population can be very laborious. It may be more convenient to carry out **systematic sampling**. This involves selecting every kth member of the population, for example, every 10th item from a particular machine on a production line.

The stratified sample can be constructed in proportion to the number of members in each stratum.

The method of **stratified sampling** is often used when the population is split into distinguishable layers or strata, such as students in each faculty in a college.

Unbiased estimates

AQA	S1
OCR	S2
CCEA	S2

When a **population parameter**, such as the mean or the variance, is **unknown**, then it is sensible to estimate it from a sample.

An **unbiased estimate** is one which, on the average, gives the true value, i.e. E(estimate) = true value of parameter.

The best unbiased estimate is the estimate with the smallest variance.

Best unbiased estimates:

For mean μ $\qquad \hat{\mu} = \bar{x} = \dfrac{\sum x}{n}$ $\qquad \bar{x}$ is sample mean

For variance σ^2 $\qquad \hat{\sigma}^2 = \dfrac{n}{n-1} s^2$ $\qquad s^2$ is the sample variance.

Key points from AS

- **Mean, variance and standard deviation**
  Revise AS pages 88–89

Make sure that you are familiar with the format given in your examination booklet and practise using it.

Sometimes S^2 is used for $\hat{\sigma}^2$

Alternative formats: $S^2 = \hat{\sigma}^2 = \dfrac{\sum(x-\bar{x})^2}{n-1}$, $S^2 = \hat{\sigma}^2 = \dfrac{1}{n-1}\left(\sum x^2 - \dfrac{(\sum x)^2}{n}\right)$.

If you are given only summary data such as Σx or Σx^2, substitute into the appropriate formula.

A calculator in statistical mode can be used to obtain the value of $\hat{\sigma}$ directly from raw data. Look for the key marked $\sigma_{x_{n-1}}$.

Sampling distributions

AQA ▶ S1
OCR ▶ S2
WJEC ▶ S2
CCEA ▶ S2

The most important sampling distribution is the distribution of the sample mean and you will need to be able to use it to find probabilities.

The sampling distribution of the mean

If all possible samples of size n are taken from $X \sim N(\mu, \sigma^2)$ and their sample means calculated, then these means form the sampling distribution of means, $\overline{X}$ which is also normally distributed.

This important result can be derived using expectation algebra (page 72).

> **KEY POINT**
>
> If $X \sim N(\mu, \sigma^2)$, then $\overline{X} \sim N\left(\mu, \dfrac{\sigma^2}{n}\right)$.

The mean of $\overline{X}$ is the same as the mean of X.

The variance of $\overline{X}$ is much smaller than the variance of X.

Standard deviation = $\sqrt{\text{variance}}$.

The diagram shows the curves for X and for $\overline{X}$.

They are both symmetrical about μ.
The curve for $\overline{X}$ is much more squashed in, confirming the smaller standard deviation.

The standard deviation of the sampling distribution, $\sigma/\sqrt{n}$, is known as the **standard error of the mean**.

The following result is extremely useful and should be learnt.

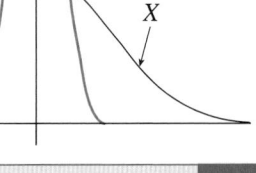

> **KEY POINT**
>
> If the distribution of X is not normal, then, by the **central limit theorem**, $\overline{X} \sim N\left(\mu, \dfrac{\sigma^2}{n}\right)$ approximately, provided n is large.

$\hat{\sigma}^2$ is the unbiased estimate of σ^2.

If you do not know σ^2, the variance of X, then it can be estimated by $\hat{\sigma}^2$ and provided n is large, $\overline{X} \sim N\left(\mu, \dfrac{\hat{\sigma}^2}{n}\right)$ approximately.

Example

The heights of men in a particular area are normally distributed with mean 176 cm and standard deviation 7 cm. A random sample of 50 men is taken. Find the probability that the mean height of the men in the sample is less than 175 cm.

If X is the height, in centimetres, of a man from this area, then $X \sim N(176, 7^2)$.

For random samples of size 50, $\overline{X} \sim N\left(176, \dfrac{7^2}{50}\right)$, i.e. $\overline{X} \sim N(176, 0.98)$.

In this example, since the population is normal, any size sample, large or small, could have been taken.

$$P(\overline{X} < 175) = P\left(Z < \frac{175 - 176}{\sqrt{0.98}}\right)$$
$$= P(Z < -1.010\ldots)$$
$$= 1 - \Phi(1.010)$$
$$= 1 - 0.8438$$
$$= 0.156 \text{ (3 d.p.)}$$

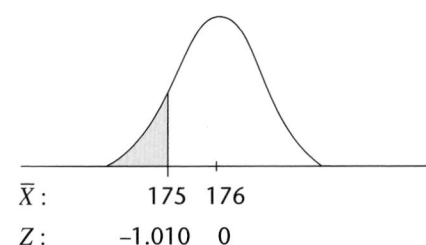

WJEC   S2

The sampling distribution of the difference between means

Taking random samples of size n_1 from $X \sim N(\mu_1, \sigma_1^2)$ and size n_2 from $Y \sim N(\mu_2, \sigma_2^2)$, the sampling distribution of the difference between means is also

X and Y must be independent.

normally distributed, where $\bar{X} - \bar{Y} \sim N\left(\mu_1 - \mu_2, \dfrac{\sigma_1^2}{n_1} + \dfrac{\sigma_2^2}{n_2}\right)$.

If X and Y have common variance σ^2, then $\bar{X} - \bar{Y} \sim N\left(\mu_1 - \mu_2, \sigma^2\left(\dfrac{1}{n_1} + \dfrac{1}{n_2}\right)\right)$.

Confidence intervals 1

AQA   S1
WJEC   S2
CCEA   S2

A **confidence interval** for an unknown population parameter is an interval (a, b) such that there is a specified probability, often 90%, 95% or 99%, that the interval contains the true value.

Confidence interval for the mean μ

This is calculated using the mean, $\bar{x}$, of a random sample of size n.
If the population is normal, the sample can be any size. If it is not normal, then n must be large, so that the Central Limit Theorem can be used.

±1.96 are the standardised z-values for the central 95% of a standard normal distribution.

The 95% **confidence limits** for μ are $\bar{x} \pm 1.96\dfrac{\sigma}{\sqrt{n}}$.

The 95% **confidence interval** is $\left(\bar{x} - 1.96\dfrac{\sigma}{\sqrt{n}}, \ x + 1.96\dfrac{\sigma}{\sqrt{n}}\right)$.

Provided n is large, $\hat{\sigma}^2$ can be used if σ^2 is unknown.

$$\bar{x}$$
$$-1.96\frac{\sigma}{\sqrt{n}} \qquad\qquad \bar{x} + 1.96\frac{\sigma}{\sqrt{n}}$$
$$\longleftarrow \quad 2 \times 1.96\frac{\sigma}{\sqrt{n}} \quad \longrightarrow$$

The **width** of the 95% confidence interval is $2 \times 1.96\dfrac{\sigma}{\sqrt{n}}$.

Different levels of confidence

±1.645 are the standardised z-values for the central 90% of a standard normal distribution.

Any level of confidence can be used. The most common are 95%, (described above), 90% and 99%.

90% confidence interval is $\left(\bar{x} - 1.645\dfrac{\sigma}{\sqrt{n}}, \ \bar{x} + 1.645\dfrac{\sigma}{\sqrt{n}}\right)$.

±2.576 are the standardised z-values for the central 99% of a standard normal distribution.

99% confidence interval is $\left(\bar{x} - 2.576\dfrac{\sigma}{\sqrt{n}}, \ \bar{x} + 2.576\dfrac{\sigma}{\sqrt{n}}\right)$.

Example

X is the mass, in grams, of a bag of flour packed by a particular machine. The mean mass of bags packed by this machine is μ and the standard deviation is σ. A random sample of 100 bags gave the following results:

$$\Sigma x = 50\,140, \ \Sigma x^2 = 25\,166\,580.$$

Find a 95% confidence interval for μ.

Since σ^2 is unknown, find $\hat{\sigma}^2$.

$$\bar{x} = \frac{\Sigma x}{n} = \frac{50\,140}{100} = 501.4$$

$$\hat{\sigma}^2 = \frac{1}{n-1}\left(\Sigma x^2 - \frac{(\Sigma x)^2}{n}\right)$$

You can work out the confidence interval straight away, but it is often easier to calculate the limits first.

$$= \frac{1}{99}\left(25\,166\,580 - \frac{50\,140^2}{100}\right) = 266.5, \text{ so } \hat{\sigma} = \sqrt{266.5} = 16.32$$

95% confidence limits are $\bar{x} \pm 1.96\,\dfrac{\hat{\sigma}}{\sqrt{n}} = 501.4 \pm 1.96\,\dfrac{16.32}{\sqrt{100}} = 501.4 \pm 3.20$

95% confidence interval = $(501.4 - 3.20,\ 501.4 + 3.20)$
$$= (498.2 \text{ g},\ 504.6 \text{ g})\ (1 \text{ d.p.})$$

The distribution of X is not known. Since n is large, the central limit theorem has been used in the underlying theory.

Confidence intervals for the difference between means

WJEC   S2

With the notation used on page 78, the confidence interval is as shown below when σ_1^2 and σ_2^2 are known.

95% confidence interval: $\left(\bar{x} - \bar{y} - 1.96\sqrt{\dfrac{\sigma_1^2}{n_1} + \dfrac{\sigma_2^2}{n_2}},\ \bar{x} - \bar{y} + 1.96\sqrt{\dfrac{\sigma_1^2}{n_1} + \dfrac{\sigma_2^2}{n_2}}\right)$

Progress check

1 The random variable X has mean μ and standard deviation σ.
   Ten independent observations of X are
   4.6, 3.9, 4.2, 5.8, 6.3, 4.2, 7.2, 6.0, 7.1, 5.4.

   (a) Find unbiased estimates of μ and σ.
   (b) Given that X is normally distributed with variance 2, calculate a 99% confidence interval for μ.

2 A random sample of size 100 is taken from a normal distribution with mean 120 and standard deviation 5.

   (a) Find P$(118.75 < \bar{X} < 120.5)$, where $\bar{X}$ is the sample mean.
   (b) State, with a reason, whether your answer would be different if X is not normally distributed.

WJEC

3 X and Y are normally distributed with variance 25.
   A sample of 10 observations from X has mean 35.2 and a sample of 20 observations from Y has mean 32.6.
   Calculate a 90% confidence interval for the difference between means.

3 $(-0.586,\ 5.786)$
2 0.8351; no, use central limit theorem since n is large
1 (a) 5.47, 1.21   (b) (4.32, 6.62)

2.4 Hypothesis tests 1

After studying this section you should be able to:

- understand the language used in hypothesis testing
- understand Type I and Type II errors
- understand how to obtain critical z-values
- perform the z-test for the mean and the difference between means
- perform tests for a binomial proportion (large and small sample sizes)
- perform a discrete test for a Poisson mean (small and large λ)
- perform a z-test for the difference between proportions (large samples)

LEARNING SUMMARY

The language of hypothesis testing

AQA	S2
EDEXCEL	S2
OCR	S2
WJEC	S2
CCEA	S2

When you want to know something about a population, you might perform a **hypothesis** or **significance test**.

A **null hypothesis**, H_0 is made about the population.
Then the **alternative hypothesis** H_1, the situation when H_0 is not true, is stated.

Example

When investigating the mean of a normal distribution, you might make the null hypothesis that the mean is 5. This is written $H_0: \mu = 5$.

The alternative hypothesis would be one of the following:

Use $\mu > 5$ if you suspect an *increase*.

Use $\mu \neq 5$ if you suspect a *change*.

- $H_1: \mu > 5$ (one-tailed, upper tail test)
- $H_1: \mu < 5$ (one-tailed, lower tail test)

Use $\mu < 5$ if you suspect a *decrease*.

- $H_1: \mu \neq 5$ (two-tailed test).

The formula for the test statistic depends on the sampling distribution. Examples are shown in the following text.

A **test statistic** is defined and its distribution when H_0 is true is stated.
Its value, based on information from a random sample taken from the population, is found. This is the **test value**.

The hypothesis test involves deciding whether or not the test value could have come from the distribution defined by the null hypothesis.
If the test value is in the main bulk of the distribution (the **acceptance region**) it is *likely* to have come from the distribution. If it is in the 'tail' end (the **rejection** or **critical region**), it is *unlikely* to have come from the distribution.

Probability theory is used to decide the placing of the boundary between 'likely' and 'unlikely'.

The decision rule (**rejection criterion**) is based on the **significance level** of the test, which is used to fix the boundaries for the rejection region. The boundaries are called **critical values**.

The significance level is a given as a percentage.

The levels used most often are 10%, 5% and 1%.

For **example**, for a significance level of 5%, the critical values are such that 5% of the distribution is in the critical (rejection) region,

There are two critical values in a two-tailed test.

i.e. $P(X > \text{critical value}) = 0.05$, for a one-tailed, upper tail test
$P(X < \text{critical value}) = 0.05$, for a one-tailed, lower tail test
$P(X > \text{upper critical value}) = 0.025$, for a two-tailed test
$P(X < \text{lower critical value}) = 0.025$, for a two-tailed test.

A test statistic is said to be significant if it lies in the critical region. If it lies in the acceptance region, then it is not significant.

Depending on the position of the test value, the **decision** is made.
If the test value lies in the critical region, H_0 is rejected in favour of H_1.
If the test value is in the acceptance region, H_0 is not rejected.

The **conclusion** is then stated in relation to the situation being tested.

Type I and Type II errors

AQA   S2
OCR   S2

In a hypothesis test, either H_0 is rejected or H_0 is not rejected.
There are occasions when the decision is incorrect and an error is made.

> You make a Type I error when you reject a true null hypothesis.

If H_0 is rejected when it is in fact true, a **Type I** error is made.

The probability of making a Type I error is the same as the significance level of the test. For example, if the significance level is 5%, then P(Type I error) = 0.05.

> You make a Type II error when you accept a false null hypothesis.

If H_0 is accepted (i.e. not rejected) when it is in fact false, a **Type II** error is made.

To find the probability of making a Type II error, a specific value for H_1 must be given. Then P(Type II error) = P(H_0 is accepted when H_1 is true).

The **power** of a test = 1 – P(Type II error).

Critical values for z-tests

AQA   S2
OCR   S2
WJEC   S2
CCEA   S2

Some hypothesis tests involving the normal distribution are known as z-tests.
In a z-test, the critical values for the rejection region can be found using the standard normal distribution, Z. The most commonly used values are summarised below.

> 5%
> For one-tailed upper tail,
> $\Phi(z) = 0.95$
> For two tailed, upper tail,
> $\Phi(z) = 0.975$
> Use symmetry for lower tails.
>
> 10%
> For one-tailed upper tail, $\Phi(z) = 0.90$
> For two tailed, upper tail $\Phi(z) = 0.95$
>
> 1%
> For one-tailed upper tail, $\Phi(z) = 0.99$
> For two tailed, upper tail $\Phi(z) = 0.995$

Level	Type of test	Critical value	Decision rule
5%	One-tailed (upper)	1.645	Reject H_0 if $z > 1.645$
	One-tailed (lower)	–1.645	Reject H_0 if $z < -1.645$
	Two-tailed	±1.96	Reject H_0 if $z > 1.96$ or $z < -1.96$
10%	One-tailed (upper)	1.282	Reject H_0 if $z > 1.282$
	One-tailed (lower)	–1.282	Reject H_0 if $z < -1.282$
	Two-tailed	±1.645	Reject H_0 if $z > 1.645$ or $z < -1.645$
1%	One-tailed (upper)	2.326	Reject H_0 if $z > 2.326$
	One-tailed (lower)	–2.326	Reject H_0 if $z < -2.326$
	Two-tailed	±2.576	Reject H_0 if $z > 2.576$ or $z < -2.576$

z-test for the mean

AQA   S2
OCR   S2
WJEC   S2
CCEA   S2

This is used to test the mean μ of a population when you know the variance σ^2.
If the population is **normal**, the samples can be **any size**, but if the population is **not normal**, then **large** samples must be taken, so that the central limit theorem can be applied.

The distribution considered for the test statistic is the sampling distribution of means (see page 77).

> When the population is not normal, the central limit theorem is needed.

> **KEY POINT**
>
> Test statistic $Z = \dfrac{\bar{X} - \mu}{\sigma/\sqrt{n}} \sim N(0, 1)$.

This formula can also be used if the variance of X is not known, provided n is large. In this case, the unbiased estimate $\hat{\sigma}^2$ is used for σ^2.

Example

A machine fills tins of baked beans such that the mass of a filled tin is normally distributed with mean mass 429 g and the standard deviation is 3 g. The machine breaks down and after being repaired, the mean mass of a random sample of 100 tins is found to be 428.45 g. Is this evidence, at the 5% level, that the mean mass of tins filled by this machine has decreased? Assume that the standard deviation remains unaltered.

Define your variables.

Let X be the mass, in grams, of a filled tin and let the mean after the repair be μ.

State the hypotheses.

$H_0: \mu = 429; H_1: \mu < 429$.

State the distribution of the test statistic assuming the null hypothesis is true.

Consider the sampling distribution of means $\overline{X} \sim N\left(429, \dfrac{3^2}{100}\right)$.

State the type of test and the rejection criterion.

Perform a one-tailed (lower tail) test at the 5% level and reject H_0 if $z < -1.645$.

Calculate the value of the test statistic.

$\overline{x} = 428.45$, so

$$z = \frac{\overline{x} - \mu}{\sigma/\sqrt{n}} = \frac{428.45 - 429}{3/\sqrt{100}} = -1.833 \ldots$$

Decide whether to reject H_0 or not and relate your conclusion to the question.

Since $z < -1.645$, the test value lies in the critical region and H_0 is rejected.

There is evidence, at the 5% level, that the mean mass of a filled tin has decreased.

Alternatively, a **probability method** can be used.
The rejection rule is to reject H_0 if $P(\overline{X} < 428.45) < 0.05$.

$$P(\overline{X} < 428.45) = P\left(Z < \frac{428.45 - 429}{3/\sqrt{100}}\right)$$

$$= P(Z < -1.833 \ldots) = 0.0334.$$

Since $P(\overline{X} < 428.45) < 0.05$, H_0 is rejected and the conclusion is as above.

z-test for difference between means

WJEC S2

This z-test is used to compare the means of two independent normal populations. A sample of size n_1 is taken from $X_1 \sim N(\mu_1, \sigma_1^2)$ and a sample of size n_2 is taken from $X_2 \sim N(\mu_2, \sigma_2^2)$.

The sampling distribution for the test statistic is the **difference between means** (page 78).

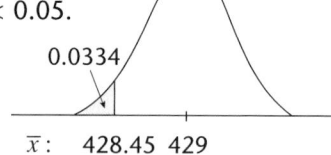

> **KEY POINT**
>
> Test statistic $Z = \dfrac{\overline{X}_1 - \overline{X}_2 - (\mu_1 - \mu_2)}{\sqrt{\dfrac{\sigma_1^2}{n_1} + \dfrac{\sigma_2^2}{n_2}}} \sim N(0, 1)$
>
> If X and Y have common variance σ^2, then
>
> $Z = \dfrac{\overline{X}_1 - \overline{X}_2 - (\mu_1 - \mu_2)}{\sigma \sqrt{\dfrac{1}{n_1} + \dfrac{1}{n_2}}} \sim N(0, 1)$

Tests for a binomial proportion

This binomial test is used to test the proportion of successes, p in a binomial population $X \sim B(n, p)$.

OCR S2
WJEC S2

Normal approximation to the binomial, see p. 75

When testing in the upper tail, subtract 0.5, and when testing in the lower tail, add 0.5.

Large sample size

When n is large (such that $np > 5$ and $nq > 5$) a normal approximation to the binomial distribution can be used.

A continuity correction of $+0.5$ or -0.5 is needed as the binomial distribution is discrete and the normal distribution is continuous.

> **KEY POINT**
>
> Test statistic $Z = \dfrac{(X \pm 0.5) - np}{\sqrt{npq}} \sim N(0, 1)$, provided n is large.

The hypothesis test follows the same pattern as the z-test for the mean.

Example

A coin is tossed 50 times and 30 heads are obtained. Is this evidence, at the 5% level of significance, that the coin is biased in favour of heads?

Define the variable.

Let the probability that the coin shows heads be p.
If X is the number of heads in 50 tosses, then $X \sim B(50, p)$.

State the hypotheses and also the distribution if the null hypothesis is true.

H_0: $p = 0.5$ (the coin is fair), H_1: $p > 0.5$ (the coin is biased in favour of heads).
If the null hypothesis is true, then $X \sim B(50, 0.5)$.

Justify the normal approximation.

But n is large such that $np = 50 \times 0.5 = 25 > 5$ and $nq = 25 > 5$, so $X \sim N(np, npq)$ approximately. Since $npq = 50 \times 0.5 \times 0.5 = 12.5$, $X \sim N(25, 12.5)$.

State the rejection criterion.

At the 5% level, reject H_0 if $z > 1.645$.

29.5 is used because you need to test whether the whole rectangle representing $x = 30$ (from 29.5 to 30.5) is in the upper tail critical region.

The test value is $x = 30$, so using the continuity correction,
$$z = \frac{(x - 0.5) - np}{\sqrt{npq}} = \frac{29.5 - 25}{\sqrt{12.5}} = 1.27 \ldots$$

Since $z < 1.645$, the test value does not lie in the critical region and H_0 is not rejected.

Relate the conclusion to the question.

At the 5% level there is not sufficient evidence to say that the coin is biased in favour of heads.

EDEXCEL S2
OCR S2
WJEC S2

Binomial test when the sample size is small

When n is small, the normal approximation does not apply, so binomial probabilities are used to decide whether the test value is in the critical region or not.

For **example**, at the 5% level, the decision rules are:

- for a one-tailed (upper tail) test, reject H_0 if $P(X \geqslant \text{test value}) < 0.05$
- for a one-tailed (lower tail) test, reject H_0 if $P(X \leqslant \text{test value}) < 0.05$
- for a two-tailed test, reject H_0 if $P(X \leqslant \text{test value}) < 0.025$ or $P(X \geqslant \text{test value}) < 0.025$.

Example

This is a small sample, so the normal approximation cannot be used.

When a die was thrown 10 times, a six occurred 4 times. Is this evidence, at the 10% level of significance, that the die is biased in favour of sixes?

Let the probability of obtaining a six be p.
If X is the number of sixes in 10 throws, then $X \sim B(10, p)$.

This is a one-tailed (upper tail) test.

H_0: $p = \frac{1}{6}$ (the die is fair), H_1: $p > \frac{1}{6}$ (the die is biased in favour of sixes).

If the null hypothesis is true, then $X \sim B(10, \frac{1}{6})$.
Reject H_0 if the test value of $x = 4$ lies in the critical region (the upper tail 10%)
i.e. reject H_0 if $P(X \geqslant 4) < 0.1$.

It may be possible to calculate the probability using cumulative binomial tables. Check in your examination booklet.

$P(X \geqslant 4) = 1 - P(X < 4)$

$$= 1 - \left(\left(\tfrac{5}{6}\right)^{10} + {}^{10}C_1\left(\tfrac{5}{6}\right)^9\left(\tfrac{1}{6}\right) + {}^{10}C_2\left(\tfrac{5}{6}\right)^8\left(\tfrac{1}{6}\right)^2 + {}^{10}C_3\left(\tfrac{5}{6}\right)^7\left(\tfrac{1}{6}\right)^3 \right)$$

$$= 1 - 0.930\ldots = 0.070 \text{ (2 s.f.)}.$$

Since $P(X \geqslant 4) < 0.1$, H_0 is rejected.

There is evidence at the 10% level, that the die is biased in favour of sixes.

Test for a Poisson mean

EDEXCEL S2 (λ small)
OCR S2 (λ small)
WJEC

The test statistic is X, the number of occurrences, where $X \sim \text{Po}(\lambda)$.

When λ is small, the test is similar to the small sample binomial test.

When λ is large, $X \sim N(\lambda, \lambda)$ approximately and the test is similar to the large sample binomial test. A continuity correction is needed.

Example

The office manager claims that the average number of telephone calls per minute received by her office switchboard is 4.5. Her supervisor maintains that it is less than 4.5. In a randomly selected minute, 2 calls were received. A hypothesis test is conducted at the 5% level. Whose claim is upheld?

Let X be the number of calls per minute. Assuming that calls occur randomly and independently, $X \sim \text{Po}(\lambda)$.

This is a one-tailed (lower tail) test.

H_0: $\lambda = 4.5$, H_1: $\lambda < 4.5$. If the null hypothesis is true, $X \sim \text{Po}(4.5)$.

The test value is $x = 2$, so, at the 5% level, reject H_0 if $P(X \leqslant 2) < 0.05$.

Since λ is small, the method is similar to the small sample binomial test.

$$P(X \leqslant 2) = e^{-4.5} + 4.5e^{-4.5} + \frac{4.5^2}{2!}e^{-4.5} = e^{-4.5}\left(1 + 4.5 + \frac{4.5^2}{2!}\right) = 0.173\ldots$$

Since $P(X \leqslant 2) > 0.05$, the test value does not lie in the critical region and H_0 is not rejected. The office manager's claim that the mean is 4.5 is upheld.

Progress check

1 $X \sim N(\mu, 4)$. The mean of a random sample of 10 items from X is 20.9. Test, at the 5% level, the hypotheses H_0: $\mu = 20$, H_1: $\mu > 20$.

2 The heights of a particular plant have mean μ and variance σ^2 (both unknown). The heights H, in centimetres, of a random sample of 100 plants are summarised as follows: $\Sigma h = 2941$, $\Sigma h^2 = 88\,502$.

(a) Find unbiased estimates for μ and σ.
(b) Does the sample data support the claim that the mean height is less than 30 cm? Perform a hypothesis test at the 10% level and state your hypotheses.

3 A random observation, x is taken from $X \sim B(n, p)$. Test, at the 5% level, the null hypothesis that $p = 0.4$, against the alternative hypothesis that $p < 0.4$.

Hint:
In (a) use small sample binomial test.
In (b) use large sample binomial test.

(a) if $n = 10$ and $x = 1$,   (b) if $n = 100$ and $x = 32$.

4 A single observation is taken from a Poisson distribution with mean λ and used to test the null hypothesis $\lambda = 7$ against the alternative hypothesis $\lambda \neq 7$. What is the conclusion if the hypothesis test is carried out at the 10% level and the observation is

Cumulative Poisson tables are needed for (b).

(a) 3   (b) 13?

1 $z = 1.423$, do not reject H_0.

2 (a) 29.41, 4.503   (b) H_0: $\mu = 30$, H_1: $\mu < 30$; $z = -1.310$, lies in critical region so reject H_0 and uphold claim.

3 (a) $P(X > 1) = 0.0464 < 0.05$, reject H_0; conclude $p < 0.4$.
(b) $z = -1.531$, $-1.645 < z$, do not reject H_0, conclude that p could be 0.4.

4 (a) $P(X > 3) < 0.05$, H_0 is not rejected.   (b) $P(X > 13) < 0.05$, H_0 is rejected.

2.5 Hypothesis tests 2

After studying this section you should be able to:

- use the t-distribution to obtain confidence intervals for the mean of a normal population with unknown variance, based on a small sample
- perform a t-test for the mean of a normal population with unknown variance, using small samples
- perform a test for the product moment correlation coefficient
- perform a χ^2 test for independence using contingency tables

LEARNING SUMMARY

The t-distribution

AQA   S2
CCEA   S2

If you take a small sample from a **normal population** with **unknown mean** μ and **unknown variance**, σ^2, the *t*-distribution is needed to obtain confidence intervals for μ and perform significance tests.

The *t*-distribution has a similar shape bell shape to the normal distribution.

> For large values of v, the *t*-distribution approaches a normal distribution.

It has one parameter, v, known as the number of **degrees of freedom**.

For a particular value of v, the appropriate *t*-distribution is denoted by $t(v)$. Tables give values of t for which $P(T < t) = p$, for various values of p, usually $p = 0.75$, 0.90, 0.95, 0.975, 0.99.

Confidence intervals 2

AQA   S2
CCEA   S2

For small samples, size n, taken from $X \sim N(\mu, \sigma^2)$, with σ^2 unknown:

> σ^2 is estimated using $\hat{\sigma}^2$ (see p. 76).

The 95% confidence limits for μ are $\bar{x} \pm t \dfrac{\hat{\sigma}}{\sqrt{n}}$,

where $(-t, t)$ encloses the central 95% of the $t(n-1)$ distribution.

> The critical *t*-value used in the confidence interval is found from *t*-distribution tables, with $v = n - 1$.

The 95% confidence interval for μ is $\left(\bar{x} - t\dfrac{\hat{\sigma}}{\sqrt{n}}, \ \bar{x} + t\dfrac{\hat{\sigma}}{\sqrt{n}}\right)$.

Example

> The population must follow a normal distribution.

A random sample of 8 observations from $X \sim N(\mu, \sigma^2)$ gave
$\Sigma x = 36.4$, $\Sigma x^2 = 188.08$

(a) Find an unbiased estimate of σ.
(b) Find a 95% confidence interval for μ.

(a) $\hat{\sigma}^2 = \dfrac{1}{n-1}\left(\Sigma x^2 - \dfrac{(\Sigma x)^2}{n}\right)$

        $= \dfrac{1}{7}\left(188.08 - \dfrac{(36.4)^2}{8}\right) = 3.208 \ldots$

     so $\hat{\sigma} = \sqrt{3.208 \ldots} = 1.791 \ldots$

(b) Since $v = 7$, use the $t(7)$ distribution.

> You need to find t such that $P(-t < T < t) = 0.95$, i.e. $P(T < t) = 0.975$.

Since the confidence interval is symmetric (two-tailed), find $v = 7$, $p = 0.975$ (2.5% in the upper tail). This gives critical value 2.365.

The 95% confidence limits for μ are $\bar{x} \pm t\dfrac{\hat{\sigma}}{\sqrt{n}} = 4.55 \pm 2.365\left(\dfrac{1.791 \ldots}{\sqrt{8}}\right)$

                                $= 4.55 \pm 1.497 \ldots$

The 95% confidence interval $= (4.55 - 1.497 \ldots, \ 4.55 + 1.497 \ldots)$

                              $= (3.05, 6.05)$ (2 d.p.)

t-tests

AQA   S2
CCEA   S2

t-test for the mean

When testing the mean from a normal population with unknown variance, when the sample size is small, the test statistic T follows a $t(n-1)$ distribution.

> The distribution of X must be normal.
>
> $\hat{\sigma}$ is the unbiased estimate of σ.

KEY POINT

Test statistic $T = \dfrac{\overline{X} - \mu}{\hat{\sigma}/\sqrt{n}} \sim t(n-1)$.

Example

Using the data given in the preceding example:
test at the 5% level the claim that μ is greater than 4.

> The null hypothesis is $\mu = 4$, even though you are asked to test the claim that $\mu > 4$.

$H_0: \mu = 4,\ H_1: \mu > 4$.

> You want t such that $P(T < t) = 0.95$.

Using t-tables for $t(7)$,
the critical value for $v = 7$, $p = 0.95$ (one-tailed test, 5% in the upper tail) is 1.895
so reject H_0 if $t > 1.895$.

$$\overline{x} = \frac{\sum x}{n} = \frac{36.4}{8} = 4.55,$$

$$\Rightarrow t = \frac{\overline{x} - \mu}{\hat{\sigma}/\sqrt{n}} = \frac{4.55 - 4}{1.791/\sqrt{8}} = 0.868 \ldots$$

Since $t < 1.895$, the test value does not lie in the critical region. There is no evidence to support the claim that μ is greater than 4.

Test for correlation coefficient

CCEA   S2

In the AS course you may have calculated the **product-moment correlation coefficient** which takes values between -1 and 1. A hypothesis test enables you to test the strength of the correlation. For example, is a coefficient of 0.6 high enough to say that that there is positive correlation?

> **Key points from AS**
>
> • **Correlation and regression**
> *Revise AS page 102*

The following notation is often used.
Product-moment coefficient: population value ρ,   sample value r

Stages in the hypothesis test for correlation coefficient

Make the **null hypothesis, H_0** that the correlation coefficient is zero.

The test is one-tailed or two-tailed, depending on the **alternative hypothesis H_1**.

$H_0: \rho = 0,\ H_1: \rho > 0$ (one-tailed, upper tail test for positive correlation).
$H_0: \rho = 0,\ H_1: \rho < 0$ (one-tailed, lower tail test for negative correlation).
$H_0: \rho = 0,\ H_1: \rho \neq 0$ (two-tailed test for correlation, positive or negative).

Decide on the **significance level**, such as 10%, 5% or 1%. This defines the **critical values** for the **rejection region**.

The critical values can be obtained from tables, for **example**, at the 5% level:

For a one-tailed test, find the value in the column headed 0.05 (5% in the tail). This is the critical value for an upper tail test. Take the negative of it for a lower tail test.

For a two-tailed test, look for the column headed 0.025 (2.5% in each tail). Take that value for the upper critical value and the negative of it for the lower critical value.

Calculate the sample correlation coefficient and compare it with the critical value.
   If sample value > critical value, reject H_0 (in the upper tail).
   If sample value < critical value, reject H_0 (in the lower tail).

Make your decision and relate it to the situation being investigated.

Example

The 10 students in a mathematics class obtained these results in their mock and actual examinations.
Find the product moment correlation coefficient and test, at the 1% level, whether there is positive correlation between the results in the two examinations.

Candidate	A	B	C	D	E	F	G	H	I	J
Mock result	45	62	85	59	27	73	74	51	72	71
Actual result	54	67	72	73	31	79	63	61	80	74

From the calculator in LR mode: $r = 0.8572$.

H_0: $\rho = 0$, H_1: $\rho > 0$, where the population correlation coefficient is ρ.

From tables, at 1% level, the critical value is 0.6851, so H_0 is rejected if $r > 0.6851$.

Since $r = 0.8572 > 0.6851$, H_0 is rejected. There is evidence, at the 1% level, of positive correlation between the results in the two examinations.

χ^2 *tests for association using contingency tables*

This is a 3 by 4 table

A χ^2 test can be used to test whether two factors are independent or whether there is an association between them.

Data are in the form of a **contingency table**, an array displaying data relating to the two factors. The size of the table is described by the number of rows and the number of columns. An h by k table has h rows and k columns.

The **observed frequencies**, O, are put into the table.

Here are the stages of the test.

- Make the **null** and **alternative hypotheses**:
  The null hypothesis, H_0, is that the two factors are independent.
  The alternative hypothesis, H_1, is that there is an association between them.

- Calculate row totals, column totals and the grand totals for the observed data and use these to calculate the **expected frequencies**, E, if the factors are independent.
  The expected frequency for a particular cell is calculated using

$$E = \frac{(\text{row total}) \times (\text{column total})}{\text{grand total}}$$

Make a table showing the expected frequencies.

- Work out v, the number of **degrees of freedom**. This depends on the size of the contingency table. For an h by k table, $v = (h-1)(k-1)$.
- Decide on the significance level of the test and look up the critical value in χ^2 tables.

- Calculate the test statistic $\chi^2 = \sum \dfrac{(O - E^2)}{E}$.

  For a 2 by 2 contingency table, where $v = 1$, it is advisable to use Yates' continuity correction. In this case, $\chi^2 = \sum \dfrac{(|O - E| - 0.5)^2}{E}$.

- Make your conclusion.
  If $\chi^2 >$ critical value, reject H_0 and conclude that the factors are not independent; there is an association between them.
  If $\chi^2 <$ critical value, do not reject H_0 and conclude that the factors are independent.

Example

Test, at the 10% significance level, whether size and colour are independent, for the data given in the following table.

> This is a 3 by 2 contingency table.

		Colour		
		Pink	White	Totals
	Small	12	23	35
Size	Medium	16	20	36
	Large	22	17	39
	Totals	50	60	110

H_0: Size and colour are independent;
H_1: Size and colour are not independent

Expected frequencies (to 1 decimal place).

Small pink: $\qquad E = \dfrac{\text{(row total)} \times \text{(column total)}}{\text{grand total}} = \dfrac{35 \times 50}{110} = 15.9$

> Only two expected frequencies need to be calculated using the formula. All the rest can then be found by ensuring that row or column totals agree. This confirms that $v = 2$.

Medium pink: $\qquad E = \dfrac{36 \times 50}{110} = 16.4$

Large pink: $\qquad E = 50 - (15.9 + 16.4) = 17.7$ (column totals argree).
Small white: $\qquad E = 35 - 15.9 = 19.1$ (row totals agree).
Medium white: $\qquad E = 36 - 16.4 = 19.6$ (row totals agree).
Large white: $\qquad E = 39 - 17.7 = 21.3$ (row totals agree).

$v = (3 - 1)(2 - 1) = 2 \times 1 = 2$. Consider the $\chi^2(2)$ distribution. At the 10% level, the critical value (from tables) is 4.605, so H_0 is rejected if $\chi^2 > 4.605$.

O	E	$\dfrac{(O - E)^2}{E}$
12	15.9	0.956 …
16	16.4	0.0097 …
22	17.7	1.044 …
23	19.1	0.796 …
20	19.6	0.00816 …
17	21.3	0.868 …
$\Sigma O = 160$	$\Sigma E = 160$	$\chi^2 = 3.6835 …$

$\chi^2 = \sum \dfrac{(O - E)^2}{E} = 3.68$

Since $\chi^2 < 4.605$, H_0 is not rejected.
Size and colour are independent.

Progress check

1 It is known that the times, in seconds, taken to perform a particular task are normally distributed with mean μ and variance σ^2.

The times, in seconds, taken by 10 volunteers are as follows:

32.9, 27.6, 33.0, 35.7, 39.2, 26.8, 34.3, 31.1, 38.9, 28.7

(a) Find an unbiased estimate of σ.

(b) Find a 99% confidence interval for μ.

(c) Find the width of the 95% confidence interval.

(d) Test at the 5% level the null hypothesis that $\mu = 35.5$, against the alternative hypothesis that $\mu < 35.5$.

2 Ten students were asked to carry out a simple mechanical task, first with their dominant hand and then with their other hand.

The times taken, in seconds, are recorded in the table.

Student	A	B	C	D	E	F	G	H	I	J
Time with dominant hand (x)	7	2	10	4	15	12	6	13	9	10
Time with other hand (y)	25	10	22	5	20	15	12	20	10	15

(a) Calculate the product-moment correlation coefficient.

(b) Test, at the 5% level, whether there is any correlation between the times taken with each hand.

3 Test, at the 5% level, whether there is an association between the two characteristics in this contingency table.

	Characteristic 2			Totals
	90	35	85	210
Characteristic 1	30	30	60	120
	30	15	25	70
Totals	150	80	170	400

1 (a) 4.37   (b) (28.3, 37.3)   (c) 6.25

(d) $t = -1.941$, reject H_0 since $t < -1.833$; mean time is less than 35.5 seconds.

2 0.395, no correlation.

3 $v = 4$, $E = 78.75$, 42, 89.25, 45, 24, 51, 26.25, 14, 29.75, $X^2 = 12.43 > 9.488$, there is an association.

Sample questions and model answers

1

The continuous random variable X has probability density function $f(x)$ given by

$$f(x) = \begin{cases} kx(2-x) & 0 \leqslant x \leqslant 2 \\ 0 & \text{otherwise} \end{cases}$$

where k is a constant.

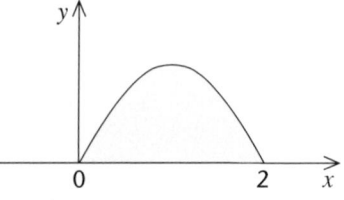

(a) Show that $k = \frac{3}{4}$ and find the exact value of $P(X > 1\frac{1}{2})$.

(b) Find $E(X)$ and $E(X^2)$.

(c) Show that the standard deviation, σ, is 0.447 correct to three decimal places.

The total area under the curve is 1.

As k is a constant, it can be taken outside the integration.

(a)
$$\int_0^2 f(x)\,dx = 1$$

$$\Rightarrow \quad k\int_0^2 (2x - x^2)\,dx = 1$$

$$\Rightarrow \quad k\left[x^2 - \frac{x^3}{3}\right]_0^2 = 1$$

$$\Rightarrow \quad k(4 - \tfrac{8}{3} - 0) = 1$$

$$\Rightarrow \quad k = \tfrac{3}{4}$$

$$f(x) = \tfrac{3}{4}(2x - x^2)$$

$$P(X > 1\tfrac{1}{2}) = \int_{1\frac{1}{2}}^2 f(x)\,dx$$

$$= \tfrac{3}{4}\int_{1\frac{1}{2}}^2 (2x - x^2)\,dx$$

Substitute the limits carefully, working in fractions to obtain the exact answer.

$$= \tfrac{3}{4}\left[x^2 - \frac{x^3}{3}\right]_{1\frac{1}{2}}^2$$

$$= \tfrac{3}{4}(4 - \tfrac{8}{3} - (\tfrac{9}{4} - \tfrac{9}{8}))$$

$$= \tfrac{5}{32}$$

If you do not spot that $f(x)$ is symmetrical about $x = 1$, find $E(X)$ by integration.

(b) By symmetry, $E(X) = 1$

$$E(X^2) = \int_0^2 x^2 f(x)\,dx$$

$$= \tfrac{3}{4}\int_0^2 (2x^3 - x^4)\,dx$$

$$= \tfrac{3}{4}\left[\frac{2x^4}{4} - \frac{x^5}{5}\right]_0^2$$

$$= \tfrac{3}{4}(8 - \tfrac{32}{5})$$

$$= 1.2$$

(c) $\text{Var}(X) = E(X^2) - [E(X)]^2$

$[E(X)]^2$ is often written $E^2(X)$.

$$= 1.2 - 1$$

$$= 0.2$$

$$\sigma = \sqrt{0.2} = 0.447 \text{ (3 d.p.)}$$

Sample questions and model answers (continued)

2

In an advanced level examination taken by a large number of candidates, the marks were distributed normally with mean mark 68.7 and standard deviation 5.4.

A random sample of 100 scripts is taken and their mean mark denoted by $\bar{X}$.

(a) State the distribution of $\bar{X}$.

(b) Find the probability that the mean mark of the 100 scripts is between 68 and 70, giving your answer correct to 2 significant figures.

Let X be the examination mark, so $X \sim N(68.7, 5.4^2)$

(a) For samples of size 100,

$$\bar{X} \sim N\left(68.7, \frac{5.4^2}{100}\right), \text{ i.e. } \bar{X} \sim N(68.7, 0.2916)$$

(b) $P(68 < \bar{X} < 70) = P\left(\dfrac{68 - 68.7}{\sqrt{0.2916}} < Z < \dfrac{70 - 68.7}{\sqrt{0.2916}}\right)$

$= P(-1.296 < Z < 2.407)$

$= \Phi(1.296) + \Phi(2.407) - 1$

$= 0.9026 + 0.9919 - 1 = 0.89$ (2 s.f.)

If the degree of approximation is not specified in the question or your examination rubric, then approximate sensibly.

3

The masses of a particular brand of buns are normally distributed with mean 50 g and standard deviation 3.4 g. They are sold in small packs of 6 or large packs of 12.

(a) Find the probability that the mass of the buns in a small pack is less than 285 g.

(b) Find the probability that the buns in two small packs weigh at least 10 g more than the buns in a large pack.

Always define the random variables carefully as this will help you to keep track of the distribution being used.

Let X be the mass of a bun. Then $X \sim N(50, 3.4^2)$.

(a) Let S be the mass of a small pack, where $S = X_1 + X_2 + \ldots + X_6$

$E(S) = 6E(X) = 300$ and $\text{Var}(S) = 6\,\text{Var}(X) = 6 \times 3.4^2 = 69.36$

$\therefore \quad S \sim N(300, 69.36)$

Note that
$S = X_1 + X_2 + \ldots + X_6$ (sum)
$S \neq 6X$ (multiple).

$P(S < 285) = P\left(Z < \dfrac{285 - 300}{\sqrt{69.36}}\right) = P(Z < -1.801) = 0.036$ (2 s.f.)

(b) Let L be the mass of a large pack, where $L = X_1 + X_2 + \ldots + X_{12}$

$E(L) = 12E(X) = 600$ and $\text{Var}(L) = 12\,\text{Var}(X) = 12 \times 3.4^2 = 138.72$

$\therefore \quad L \sim N(600, 138.72)$.

$P(S_1 + S_2 \geqslant L + 10) = P(S_1 + S_2 - L \geqslant 10)$

Let $D = S_1 + S_2 - L$

Remember the + sign in the variance.

$E(D) = 300 + 300 - 600 = 0$ and $\text{Var}(D) = 69.36 + 69.36 + 138.72 = 277.44$

$\therefore \quad D \sim N(0, 277.44)$

$P(D \geqslant 10) = P\left(Z \geqslant \dfrac{10 - 0}{\sqrt{277.44}}\right) = P(Z \geqslant 0.6003 \ldots) = 0.27$ (2 s.f.)

Sample questions and model answers (continued)

4

The lengths of a population of snakes are normally distributed with standard deviation 8 cm and unknown mean μ cm.

A random sample of 10 snakes is taken and their mean length calculated.

When a significance test, at the 10% level, is performed, the hypothesis that μ is 37.5 is rejected in favour of the hypothesis that it is greater than 37.5 cm.

(a) What can be said about the value of the sample mean?

(b) Explain briefly what is meant, in the context of this question, by a Type I error, and state the probability of making a Type I error.

(c) Find the probability of making a Type II error when $\mu = 41.1$.

(a) If X is the length, in centimetres, then $X \sim N(\mu, 8^2)$.

$H_0: \mu = 37.5$

$H_1: \mu > 37.5$

For the significance test, the sampling distribution of means is considered.

If $\mu = 37.5$, then $\bar{X} \sim N\left(37.5, \dfrac{8^2}{10}\right)$.

From standard normal tables, $\Phi(z) = 0.9 \Rightarrow z = 1.282$.

The critical value for a one-tailed (upper tail) at the 10% level is 1.282.

H_0 is rejected if z is in the upper tail 10% of the distribution.

Since H_0 is rejected, $z > 1.282$,

Test statistic is $Z = \dfrac{\bar{X} - \mu}{\sigma/\sqrt{n}}$.

i.e. $\dfrac{\bar{x} - 37.5}{8/\sqrt{10}} > 1.282$

$\bar{x} > 40.74$

$Z:$ $\quad 0 \quad 1.282$

$\quad 37.5 \quad 40.74$

10%

(b) A Type I error is made when the null hypothesis, $\mu = 37.5$, is rejected, when the mean population length is in fact 37.5.

P(Type I error) = 10%

The probability of making a Type I error is the same as the level of significance of the test.

(c) $H_0: \mu = 37.5$

$H_1: \mu = 41.1$

P(Type II error) = P(H_0 is accepted | H_1 is true)

From (a), H_0 is accepted if $\bar{x} < 40.74$.

If H_1 is true, $\bar{X} \sim N\left(41.1, \dfrac{8^2}{10}\right)$

$$\text{P(Type II error)} = P\left(\bar{X} < 40.74 \,\middle|\, \bar{X} \sim N\left(41.1, \dfrac{8^2}{10}\right)\right)$$

$$= P\left(Z < \dfrac{40.74 - 41.1}{8/\sqrt{10}}\right)$$

$$= P(Z < -0.142)$$

$$= 1 - 0.5565$$

$$= 0.44 \ (2 \text{ s.f.})$$

P (Type II error)

$\bar{X}:$ $\quad 40.74 \ 41.1$

Practice examination questions

1 It is given that $X \sim N(\mu, 16)$.

The null hypothesis $\mu = 20$ is to be tested against the alternative hypothesis $\mu \neq 20$.

The mean of a random sample of 10 observations from X is 17.2.

Test at the 5% level whether this provides evidence that μ is not 20.

2 The probability that a person suffers from a particular health condition is 0.01.

(a) Using a suitable approximation, find the probability that in a randomly chosen sample of 80 people, fewer than 3 suffer from the health condition.

(b) Find the minimum sample size in order that the probability of including at least one with the health condition is greater than 95%.

3 The continuous random variable X is distributed uniformly in the interval $2 \leqslant x \leqslant 12$.

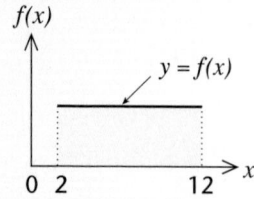

(a) Find the probability density function, $f(x)$.

(b) Calculate the probability that X lies within one standard deviation of the mean, giving your answer correct to 2 decimal places.

4 The random variable X has a normal distribution with mean μ and variance σ^2.
In order to set up a confidence interval for μ, a random sample of 9 observations of X is taken.

The results are summarised as follows:

$$\Sigma x = 255.6 \qquad \Sigma x^2 = 7372.26$$

(a) If $\sigma = 4$, calculate a 90% confidence interval for μ.

(b) If σ is unknown,

    (i) calculate an unbiased estimate for σ,
    (ii) calculate a 95% confidence interval for μ.

5 The random variable X has a binomial distribution with $n = 10$ and $p = 0.4$.
The mean of 60 random observations of X is $\bar{X}$.

Find $P(\bar{X} < 3.5)$, explaining what part the central limit theorem has played in your answer.

Practice examination questions (continued)

6 A beetle infestation is discovered in the trees in a small copse. If more than 35% of the trees are infected it is impractical to treat the trees with chemicals and they will have to be felled.

BINOMIAL TEST
– small sample

(a) The representative from the ministry checked a random sample of 10 trees and found that 6 were infected. Stating any assumptions, use a hypothesis test, at the 10% significance level, to decide whether the infestation should be treated with chemicals or the trees in the copse felled.

BINOMIAL TEST
– large sample

(b) The estate manager checked a random sample of 30 trees and found that 14 were infected. Using a suitable approximation, carry out a significance test, again at the 10% level. On the basis of this sample, how should the infestation be dealt with?

7 The continuous random variable X has probability function $f(x)$ where $f(x) = k(x - x^3)$, $0 \leqslant x \leqslant 1$ and $f(x) = 0$ otherwise.

(a) Show that $k = 4$.

(b) Find $E(X)$.

(c) Find $P(X < 0.5)$.

(d) Find $Var(X)$.

8 The continuous random variable X has cumulative distribution function given by

$$F(x) = \begin{cases} 0 & x \leqslant 0 \\ 2cx & 0 \leqslant x \leqslant 1 \\ c(4x - x^2 - 1) & 1 \leqslant x \leqslant 2 \\ 1 & x \geqslant 2 \end{cases}$$

(a) Show that $c = \frac{1}{3}$.

(b) Show that $P(X > 1\frac{1}{2}) = \frac{1}{12}$.

(c) Find the probability function $f(x)$.

(d) Calculate $E(X)$.

9 A psychologist is investigating whether or not there is any association between whether a person is left- or right-handed and the ability to complete a task involving manual dexterity in a given time. A sample of 200 people gave the following results:

	Completed task	*Did not complete task*
Right-handed	83	67
Left-handed	37	13

Stating your hypotheses clearly, test, at the 5% level, whether or not there is any evidence of an association between right- or left-handedness and the ability to complete the task in the given time.

Chapter 3
Mechanics 2

The following topics are covered in this chapter:

- Projectiles
- Linear motion with variable acceleration
- Variable acceleration using vectors
- Equilibrium of a rigid body
- Centre of mass

- Collisions and impulse
- Uniform circular motion
- Further circular motion
- Work, energy and power
- Elastic springs and strings

3.1 Projectiles

After studying this section you should be able to:

- model the motion of a projectile moving under constant acceleration
- understand the limitations of the model
- solve problems by considering, separately, the vertical and horizontal components of velocity
- derive and apply the trajectory formula

LEARNING SUMMARY

The model

AQA	M1
EDEXCEL	M2
OCR	M2
WJEC	M2
CCEA	M2

> Check the value of g used on your exam paper. Make sure that you use the given value or you may lose marks.

In the usual **particle** model for projectiles, air resistance is ignored and the only force taken to act on the particle is its weight which is constant. According to this model:

> **KEY POINT**
>
> - The horizontal component of velocity remains constant throughout the motion.
> - The vertical component of velocity is subject to a constant downward acceleration of g ms^{-2}.

The accuracy of the model, in predicting the motion of a projectile, depends on the extent to which the initial assumptions are satisfied.

For example, the model will provide accurate information about a projectile such as a stone or a dart over a short distance, but will give poor results for light objects that are affected more by air resistance.

> Be careful about using formulae for time of flight, greatest height, range etc. without justification or, again, marks may be lost.

When using the model, a first step is usually to resolve the velocity of a projectile into horizontal and vertical components. The two parts are then treated separately.

A particle moving with speed v at angle θ to the horizontal has vertical component $v \sin \theta$ and horizontal component $v \cos \theta$.

> The velocity is usually shown on a diagram as in Figure 1 but is used in equations as in Figure 2.

If you know the horizontal and vertical components of a velocity then you can combine them to find its magnitude (speed) and direction.

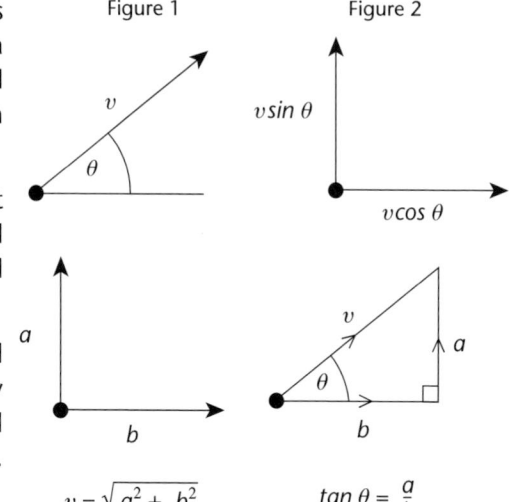

Figure 1   Figure 2

$$v = \sqrt{a^2 + b^2} \qquad \tan \theta = \frac{a}{b}$$

Upwards is taken to be positive so the acceleration is shown as negative.

Example

A stone is thrown with speed 15 ms⁻¹ at an angle of 60° above the horizontal. Find its speed and direction after 1 second. Take $g = 9.8$ ms⁻².

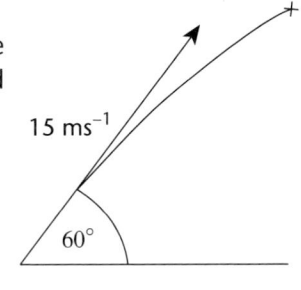

Vertically: using $v = u + at$

$$v_a = 15 \sin 60^0 - 9.8$$

$$= 3.190\ 38\ ...$$

Horizontally: $v_b = 15 \cos 60^0$

$$= 7.5$$

$$v = \sqrt{7.5^2 + 3.190\ 38^2} = 8.15 \text{ to 3 s.f.}$$

$$\tan \theta = \frac{3.19038}{7.5} \Rightarrow \theta = 23.0^0 \text{ to 1 d.p.}$$

After 1 second the stone has speed 8.15 ms⁻¹ at 23⁰ above the horizontal.

The trajectory formula

OCR   M2
WJEC   M2
CCEA   M2

Taking the point of projection as the origin, the position of a projectile may be described in terms of (x, y) coordinates.

Horizontally: $x = v \cos \theta \times t \Rightarrow t = \dfrac{x}{v \cos \theta}$ (1)

Vertically: $\quad y = v \sin \theta \times t - \frac{1}{2} g t^2$ (2)

This is the Cartesian equation of the path of the projectile. It is known as the trajectory formula.

Substituting for t in (2) gives: $y = v \sin \theta \times \dfrac{x}{v \cos \theta} - \dfrac{g x^2}{2 v^2 \cos^2 \theta}$

so: $\quad y = x \tan \theta - \dfrac{g x^2}{2 v^2}(1 + \tan^2 \theta)$

Note that:
$$\frac{1}{\cos^2 \theta} \equiv \sec^2 \theta \equiv (1 + \tan^2 \theta).$$

Example

A ball is kicked with speed 20 ms⁻¹ at 30° above the horizontal towards a wall 5 m high. The wall is 15 m from the point where the ball is kicked. Take $g = 9.8$ ms⁻². Will the ball hit the wall or clear it? Consider the effect of any assumptions.

Using the formula: $y = x \tan \theta - \dfrac{g x^2}{2 v^2}(1 + \tan^2 \theta)$

$$y = 15 \tan 30^\circ - \frac{9.8 \times 15^2}{2 \times 20^2}(1 + \tan^2 30^\circ)$$

$$= 4.99 \text{ m to 3 s.f.}$$

According to the model, the ball will hit the wall close to the top.
If the size of the ball is taken into account along with the effect of air resistance then it is clear that the ball cannot clear the wall.

Progress check

A small object is projected from ground level with speed 30 ms⁻¹ at 40° above the horizontal. Take $g = 10$ ms⁻².

(a) Find the speed and direction of the object after 1.5 seconds.

(b) The object just clears a tree 25 m from the point of projection. How high is the tree?

(a) 23.4 ms⁻¹, 10.6° above horiz (b) 15.2 m

3.2 Linear motion with variable acceleration

After studying this section you should be able to:

- set up and solve a differential equation representing acceleration in a straight line, where the acceleration is given as a function of displacement or time
- solve problems which can be modelled as the linear motion of a particle under the action of a variable force

Kinematics

AQA — M2
EDEXCEL — M2
WJEC — M2
CCEA — M1/M2

Kinematics is the branch of mathematics that deals with the motion of a particle in terms of displacement, velocity and acceleration, without considering the forces that may be required to cause the motion.

For motion in a straight line, the distinction between **distance** and **displacement** is only that displacement may be positive or negative to indicate a sense of direction, whereas distance is always taken to be positive.

The same distinction applies to **speed** and **velocity**. Velocity in a straight line may be positive or negative, depending on the direction of movement, but speed is always taken to be positive.

In the standard notation:

- displacement is represented by x
- velocity is represented by $v = \dot{x} = \dfrac{dx}{dt}$.

You may need to set up and solve equations of the form $\dfrac{dx}{dt} = f(x)$ or $\dfrac{dx}{dt} = f(t)$.

- **Acceleration** is represented by $\ddot{x} = \dfrac{dv}{dt} = v\dfrac{dv}{dx}$.

> This is an important result and you may need to establish it in the exam.

You should recognise that by the chain rule for differentiation: $\dfrac{dv}{dt} = \dfrac{dv}{dx} \times \dfrac{dx}{dt} = v\dfrac{dv}{dx}$.

> Not needed in WJEC

You may need to set up and solve equations of the form $\dfrac{dv}{dt} = f(t)$ or $v\dfrac{dv}{dx} = f(x)$.

> This requires a differential equation because the acceleration is not constant.

Example

A particle P moves in the direction of the positive x-axis with acceleration $3\sqrt{x}$ ms^{-2} directed away from O, where OP $= x$ metres. When $x = 1$ the speed of P is 2 ms^{-1}. Find its speed when $x = 4$.

> See **separation of variables** page 52.

The acceleration is $v\dfrac{dv}{dx} = 3x^{\frac{1}{2}} \Rightarrow \displaystyle\int v\,dv = \int 3x^{\frac{1}{2}}\,dx$

This gives: $\dfrac{v^2}{2} = 2x^{\frac{3}{2}} + c$

When $x = 1$, $v = 2$ so $2 = 2 + c \Rightarrow c = 0$.

so: $v^2 = 4x^{\frac{3}{2}}$

When $x = 4$, $v^2 = 32 \Rightarrow v = 4\sqrt{2}$

The particle has speed $4\sqrt{2}$ ms^{-1} when $x = 4$.

Dynamics

AQA   M2
CCEA   M2

Dynamics is concerned with the motion of a particle in response to forces that act upon it. The first step is to identify all of the forces and then use $F = ma$ to set up a differential equation.

Example

A particle P of mass 0.2 kg is attached to one end of a light elastic string. The other end of the string is attached to a fixed point on a smooth horizontal table. P is released from rest and moves along the surface of the table towards a point O.

The tension in the string has magnitude $25x$ N where x metres is the displacement of P from O. Given that $x = 2$ at the point of release, find the speed of P when $x = 1$.

Draw a diagram and label the forces.

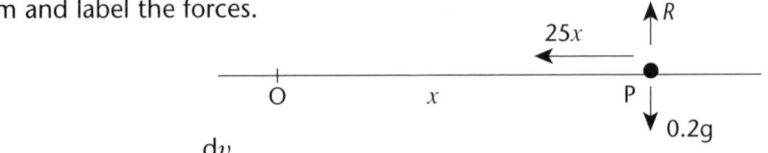

In the diagram, x increases from left to right so take this as the positive direction.

Using $F = ma$: $-25x = 0.2v\dfrac{\mathrm{d}v}{\mathrm{d}x}$

so: $\displaystyle\int -125x\,\mathrm{d}x = \int v\,\mathrm{d}v$

giving: $\dfrac{v^2}{2} = -125\dfrac{x^2}{2} + c$

Simplifying the equation can make it easier to substitute particular values.

This may be written as $v^2 = A - 125x^2$

When $x = 2$, $v = 0$ so $0 = A - 125 \times 4 \Rightarrow A = 500$

When $x = 1$, $v^2 = 500 - 125 = 375$

giving: $v = \sqrt{375}$

The speed of P when $x = 1$ is 19.4 ms^{-1} to 3 s.f.

Progress check

1 A particle P of mass 0.4 kg moves in the direction of the positive x-axis with acceleration $\dfrac{5}{x}$ ms^{-2} away from O, where OP $= x$ metres. When $x = 1$ the speed of P is 3 ms^{-1}. Find its speed when $x = 5$.

2 A particle P of mass 0.5 kg moves in a straight line, away from a point O, under the action of a force of magnitude $\dfrac{x^2}{12}$ N directed towards O, where OP $= x$ metres. When $x = 3$ the speed of P is 15 ms^{-1}. Find x when P is brought to rest.

2 12.7 m to 3 s.f.
1 5.01 ms^{-1} to 3 s.f.

3.3 *Variable acceleration using vectors*

After studying this section you should be able to:

- differentiate and integrate a vector with respect to time
- use differentiation and integration of vectors to solve problems involving variable acceleration

LEARNING SUMMARY

Vector and scalar quantities

AQA	M2
EDEXCEL	M2
WJEC	M2
CCEA	M2

Key points from AS

- **Scalars and vectors**
  Revise AS page 63

> A **scalar** quantity has **magnitude** i.e. size, but not direction. Numbers are scalars and some other important examples are distance, speed, mass and time.
>
> A **vector** quantity has both magnitude and direction. For example, distance in a specified direction is called displacement. Some other important examples are velocity, acceleration, force and momentum.
>
> **KEY POINT**

The magnitude of a vector $\mathbf{r} = a\mathbf{i} + b\mathbf{j}$ is given by $r = \sqrt{a^2 + b^2}$

Differentiation of vectors

AQA	M2
EDEXCEL	M2
WJEC	M2
CCEA	M2

A vector of the form $\mathbf{r} = f(t)\mathbf{i} + g(t)\mathbf{j}$ may be differentiated with respect to time to give $\dot{\mathbf{r}} = f'(t)\mathbf{i} + g'(t)\mathbf{j}$

For example, if $\mathbf{r} = 3t^2\mathbf{i} + \sqrt{t}\mathbf{j}$ then $\dot{\mathbf{r}} = 6t\mathbf{i} + \dfrac{1}{2\sqrt{t}}\mathbf{j}$.

> Differentiate the coefficients of **i** and **j** with respect to *t*.

If $\mathbf{r}$ represents the position vector of some moving point at time t then $\dot{\mathbf{r}}$ represents the corresponding velocity at time t.

$\dot{\mathbf{r}}$ may be differentiated to give $\ddot{\mathbf{r}}$ which represents the acceleration of the moving point at time t.

Example

A particle P has position vector $\mathbf{r}$ metres relative to a fixed point O, at time t seconds, where $\mathbf{r} = (2t^2 + 5t)\mathbf{i} + t^3\mathbf{j}$ and $t \geqslant 0$.

Find:

(a) the velocity of P at time t seconds

(b) the speed of P when $t = 3$

(c) the time when P is moving parallel to the vector $\mathbf{i} + 3\mathbf{j}$.

(a) The velocity of P in metres per second at time t is given by $\dot{\mathbf{r}} = (4t + 5)\mathbf{i} + 3t^2\mathbf{j}$.

(b) When $t = 3$, $\dot{\mathbf{r}} = 17\mathbf{i} + 27\mathbf{j}$.

> The speed of P is the magnitude of its velocity.

The speed of P when $t = 3$ is $\sqrt{17^2 + 27^2}$ ms^{-1} = 31.9 ms^{-1} to 3 s.f.

(c) When P is moving parallel to $\mathbf{i} + 3\mathbf{j}$ the component of velocity in the $\mathbf{j}$ direction must be 3 times the component in the $\mathbf{i}$ direction.

This gives:

$$3t^2 = 3(4t + 5)$$

so:

$$t^2 = 4t + 5$$

$$t^2 - 4t - 5 = 0$$

$$(t - 5)(t + 1) = 0$$

$$t = 5 \text{ or } t = -1$$

However, $t \geqslant 0$ so the only solution is $t = 5$.

Integration of vectors

AQA — M2
EDEXCEL — M2
WJEC — M2
CCEA — M2

In general, $\int (f(t)\mathbf{i} + g(t)\mathbf{j})dt = \left(\int f(t)dt\right)\mathbf{i} + \left(\int g(t)dt\right)\mathbf{j}$

> Don't be put off by the complicated looking formula, just look at what it means.

So, to integrate a vector with respect to t, integrate the coefficients of $\mathbf{i}$ and $\mathbf{j}$ with respect to t. **Note that a constant of integration is needed and that this is itself a vector.**

Example

Find $\mathbf{v} = \int \mathbf{a}\, dt$ given that $\mathbf{a} = 2t\mathbf{i} - 6t^2\mathbf{j}$ and that $\mathbf{v} = 3\mathbf{i} + \mathbf{j}$ when $t = 1$.

$\mathbf{v} = \int (2t\mathbf{i} - 6t^2\mathbf{j})dt = t^2\mathbf{i} - 2t^3\mathbf{j} + \mathbf{c}$

When $t = 1$: $\mathbf{v} = \mathbf{i} - 2\mathbf{j} + \mathbf{c} = 3\mathbf{i} + \mathbf{j}$

so: $\mathbf{c} = 2\mathbf{i} + 3\mathbf{j}$

This gives: $\mathbf{v} = t^2\mathbf{i} - 2t^3\mathbf{j} + 2\mathbf{i} + 3\mathbf{j}$
or $\mathbf{v} = (t^2 + 2)\mathbf{i} + (3 - 2t^3)\mathbf{j}$

In the example, if $\mathbf{a}$ represents acceleration then $\mathbf{v}$ represents velocity.

> **KEY POINT**
>
> Vectors for position $\mathbf{r}$, velocity $\mathbf{v}$ and acceleration $\mathbf{a}$ are linked through differentiation and integration.
>
> $\mathbf{v} = \dot{\mathbf{r}}$ $\mathbf{v} = \int \mathbf{a}\, dt$ differentiate
>
> $ \mathbf{r} \quad \mathbf{v} \quad \mathbf{a}$
>
> $\mathbf{a} = \dot{\mathbf{v}} = \ddot{\mathbf{r}}$ $\mathbf{r} = \int \mathbf{v}\, dt$ integrate

> 'Initially' means when $t = 0$.

Example

A particle P, initially at rest, has acceleration $\mathbf{a} = 2t\mathbf{i} + 3\sqrt{t}\mathbf{j}$ ms^{-2} at time t seconds.

(a) Find the velocity of P at time t seconds.
(b) Show that P is moving parallel to $\mathbf{i} + \mathbf{j}$ when $t = 4$.

> Note that the formulae for constant acceleration cannot be used here.
>
> $3 \times t^{\frac{1}{2}} \times \frac{2}{3} = 2t^{\frac{3}{2}}$.

(a) $\mathbf{v} = \int \mathbf{a}\, dt = \int (2t\mathbf{i} + 3\sqrt{t}\mathbf{j})dt$

$\phantom{(a) \mathbf{v}} = t^2\mathbf{i} + 2t^{\frac{3}{2}}\mathbf{j} + \mathbf{c}$

When $t = 0$, $\mathbf{v} = 0 \Rightarrow \mathbf{c} = 0$ so, $\mathbf{v} = t^2\mathbf{i} + 2t^{\frac{3}{2}}\mathbf{j}$

(b) When $t = 4$, $\mathbf{v} = 16\mathbf{i} + 16\mathbf{j}$
$ = 16(\mathbf{i} + \mathbf{j})$ so P is moving parallel to $\mathbf{i} + \mathbf{j}$.

Progress check

1 An object has position vector $\mathbf{r} = (2t^3\mathbf{i} - 5t^2\mathbf{j})$ m at time t seconds.
Find the acceleration of the object when $t = 3$.

2 A particle P, initially at the point with position vector $(-3\mathbf{i} + 5\mathbf{j})$ m, has velocity $\mathbf{v} = \sqrt{t}\mathbf{i} + 3\mathbf{j}$ at time t seconds. Find the position vector of P when $t = 4$.

$\text{1 } \ddot{\mathbf{r}} = 36\mathbf{i} - 10\mathbf{j} \quad \text{2 } 5\frac{1}{3}\mathbf{i} + 17\mathbf{j}$

3.4 Equilibrium of a rigid body

After studying this section you should be able to:

- *understand the conditions for equilibrium of a rigid body under the action of coplanar forces*
- *solve problems involving coplanar forces*

Equilibrium

AQA	M2
EDEXCEL	M2
OCR	M2
WJEC	M1
CCEA	M1

Forces that act in the same plane are coplanar.

You only need to establish this for *one* point and you can choose any point that is convenient.

You may need to show that a rigid body is in **equilibrium** under a system of **coplanar** forces *or* to *use* the fact that a rigid body is in equilibrium to find something about the forces acting on it.

To show that a rigid body is in equilibrium under the action of coplanar forces you need to establish that:

- the vector sum of the forces is zero
- the sum of the moments about some point is zero.

If a body is in equilibrium under some coplanar forces then you know that:

- the vector sum of the forces will be zero
- the sum of the moments about any point you choose will be zero.

> **KEY POINT**
>
> The conditions for equilibrium can be used to produce equations. The equations can then be solved to find the unknown values in a problem.

Example

A uniform plank AB of length 5 m and mass 20 kg rests horizontally on two supports, one at each end. A mass of 10 kg is positioned on the plank 1 m from B. Find the reaction at each of the supports. Take $g = 9.8$ ms^{-2}.

The first step is to draw a clearly labelled diagram.

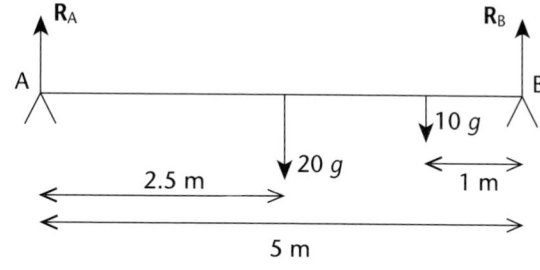

The plank is uniform so its weight acts at its mid-point.

The vector sum of the forces is zero so the total upward force must equal the total downward force.

Vertically:
$$\mathbf{R}_A + \mathbf{R}_B = 20g + 10g$$
$$= 30g$$

M(A) gives:
$$5\mathbf{R}_B = 20g \times 2.5 + 10g \times 4$$
$$= 90g$$
$$\mathbf{R}_B = 18g$$
$$= 176.4 \text{ N (reaction at B)}$$

Substitute for $\mathbf{R}_B$:
$$\mathbf{R}_A + 18g = 30g$$
$$\mathbf{R}_A = 12g$$
$$= 117.6 \text{ N (reaction at A)}.$$

Leaning ladders

AQA   M2
EDEXCEL   M2
OCR   M2

The situation where a ladder leans against a wall, possibly supporting a load at some point along its length, provides a rich source of questions on equilibrium. It requires forces and moments to be considered and includes the theory of friction.

Example

A uniform ladder of mass m kg rests against a smooth vertical wall with its lower end on rough horizontal ground. A man of mass $4m$ kg has climbed three-quarters of the way up the ladder but cannot go any higher or it will slip. The ladder makes an angle of $60°$ with the horizontal.

Find the coefficient of friction between the ladder and the ground.

Start by drawing a clearly labelled diagram.

The wall is smooth so there is no vertical force at the top of the ladder.

The length of the ladder has been written as $4l$ to simplify taking moments.

Friction acts at the bottom of the ladder to oppose the tendency for it to slip.

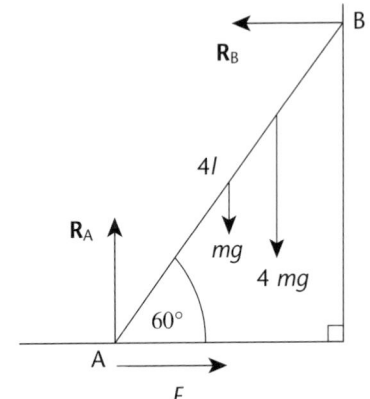

In this example, three equations may be formed by resolving the forces in two directions and taking moments.

The equations provide enough information to solve the problem.

Vertically:
$$\mathbf{R}_A = mg + 4mg \qquad (1)$$
$$= 5mg$$

Horizontally:
$$F = \mathbf{R}_B \qquad (2)$$

M(A):
$$\mathbf{R}_B \times 4l \sin 60° = mg \times 2l \cos 60° + 4mg \times 3l \cos 60°$$

giving:
$$\mathbf{R}_B \times 2\sqrt{3}l = mgl + 6mgl$$

so:
$$\mathbf{R}_B = \frac{7mg}{2\sqrt{3}} \qquad (3)$$

From (2) and (3)
$$F = \frac{7mg}{2\sqrt{3}}$$

It's a good idea to rationalise the denominator by multiplying the top and bottom of the fraction by $\sqrt{3}$.

Since the ladder is on the point of slipping, friction is limiting and the coefficient of friction is given by:
$$\mu = \frac{F}{\mathbf{R}_A} = \frac{7}{10\sqrt{3}} = \frac{7\sqrt{3}}{30}$$

Progress check

1. A uniform plank AB of length 3 m and mass 25 kg rests horizontally on two supports, one at each end. A mass of 15 kg is positioned on the plank 1 m from B. Find the reaction at each of the supports. Take $g = 9.8 \text{ ms}^{-2}$.

2. A uniform ladder of length 5 m rests against a smooth wall with its base on rough horizontal ground. The coefficient of friction between the ladder and the ground is $\frac{2}{3}$ and the ladder is on the point of slipping. Show that the angle between the ladder and the ground is $\tan^{-1}(\frac{3}{4})$.

1 220.5 N (A), 171.5 N(B)

3.5 Centre of mass

After studying this section you should be able to:

- find the centre of mass of a system of particles in one and two dimensions
- find the centre of mass of a lamina by considering an equivalent system of particles
- use centre of mass to determine conditions for the equilibrium of a lamina

Centre of mass of a system of particles

AQA	M2
EDEXCEL	M2
OCR	M2
WJEC	M1

In the diagram, m_1 and m_2 lie on a straight line through O. Their displacements from O are x_1 and x_2 respectively.

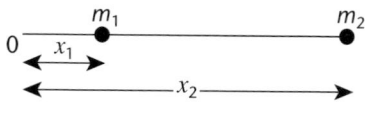

The position given by $\bar{x} = \dfrac{m_1 x_1 + m_2 x_2}{m_1 + m_2}$ is called the **centre of mass** of m_1 and m_2.

> If m_1 and m_2 are equal then this result gives the mid-point of the two masses.

For n separate masses m_1, m_2, ... m_n the position of the centre of mass is given by

$$\bar{x} = \frac{m_1 x_1 + m_2 x_2 + ... m_n x_n}{m_1 + m_2 + ... m_n}$$ This is usually written as $$\bar{x} = \frac{\sum\limits_{i=1}^{n} m_i x_i}{\sum\limits_{i=1}^{n} m_i}$$

In two dimensions, using the usual Cartesian coordinates, the position of the centre of mass is at $(\bar{x}, \bar{y})$ where $\bar{x}$ and $\bar{y}$ are given by

$$\bar{x} = \frac{\sum\limits_{i=1}^{n} m_i x_i}{\sum\limits_{i=1}^{n} m_i} \text{ as above, and } \bar{y} = \frac{\sum\limits_{i=1}^{n} m_i y_i}{\sum\limits_{i=1}^{n} m_i}$$

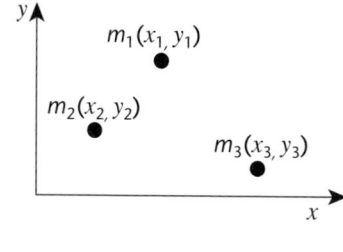

Example

Masses of 2 kg, 3 kg and 5 kg are positioned as shown in the diagram.
Find the coordinates of the centre of mass of the system.

> The method is easily extended to deal with any number of masses.

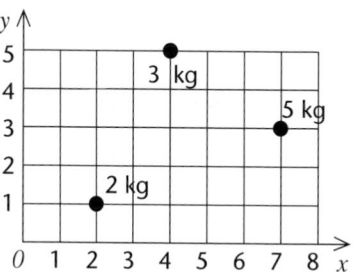

$$\bar{x} = \frac{2 \times 2 + 3 \times 4 + 5 \times 7}{2 + 3 + 5} = \frac{51}{10} = 5.1$$

$$\bar{y} = \frac{2 \times 1 + 3 \times 5 + 5 \times 3}{2 + 3 + 5} = \frac{32}{10} = 3.2$$

The centre of mass has coordinates (5.1, 3.2)

Centre of mass of a lamina

AQA	M2
EDEXCEL	M2
OCR	M2
WJEC	M1

A **lamina** is something that is thin and flat such as a sheet of metal. The thickness of a lamina is ignored and it is treated as a two-dimensional object.

You can find the centre of mass of a uniform lamina by dividing it into parts which you then represent as particles. Each part is usually rectangular or triangular and in order to represent it as a particle:

- the mass of the particle is represented by the area of the part
- the position of the particle is taken to be at the geometric centre of the part.

This representation of a lamina gives you the information you need to apply the formulae for its centre of mass.

Example

Find the coordinates of the centre of mass of this uniform lamina.

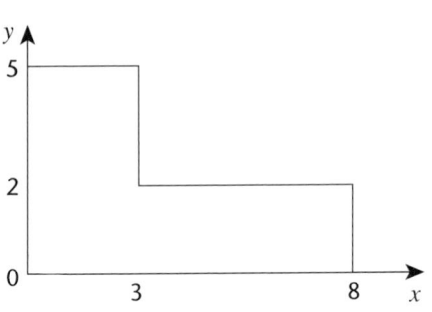

There are different ways to do this, but it doesn't matter which you choose.

The first step is to divide the lamina into rectangles.

Rectangle A has area 15 square units. Its centre is at (1.5, 2.5).

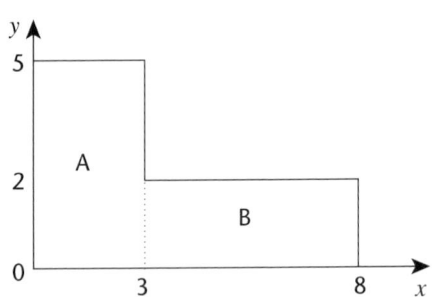

This process is easily extended to deal with more complex composite figures.

Rectangle B has area 10 square units. Its centre is at (5.5, 1).

$$\bar{x} = \frac{15 \times 1.5 + 10 \times 5.5}{15 + 10} = 3.1$$

$$\bar{y} = \frac{15 \times 2.5 + 10 \times 1}{15 + 10} = 1.9$$

The centre of mass of the lamina has coordinates (3.1, 1.9).

Freely suspended lamina

AQA	M2
EDEXCEL	M2
OCR	M2
WJEC	M1

Once you know where the centre of mass of a lamina is, you can use this to work out how it will move if it is freely suspended from a given point.

> When a lamina is freely suspended, it will move so that its centre of mass lies directly below the point of suspension.

KEY POINT

Example

The diagram shows the lamina of the previous example hanging freely from one corner.

Find the size of angle θ between the direction of Oy and the horizontal.

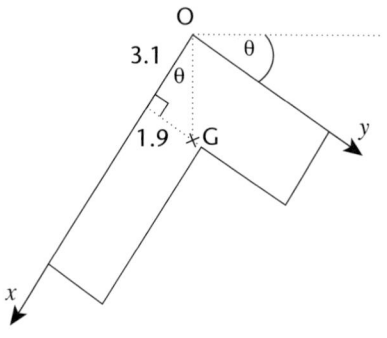

The position of the centre of mass is shown at the point G.
Notice that θ also lies in a right-angled triangle in which two of the sides are known.

From the diagram: $\tan \theta = \dfrac{1.9}{3.1}$

giving $\theta = 31.5°$ to 1 d.p.

Progress check

Masses of 2 kg, 3 kg and 5 kg are positioned as shown in the diagram.

(a) Find the coordinates of G, the centre of mass of the system.
(b) Find the angle θ between OG and Ox.

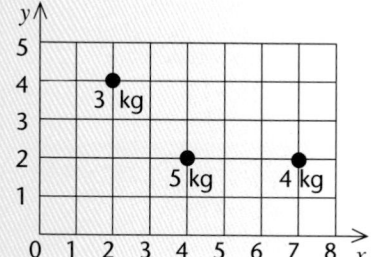

(a) G(4.5, 2.5) (b) 29.1°

3.6 Collisions and impulse

After studying this section you should be able to:

- *apply the principle of conservation of momentum to direct impact*
- *apply Newton's experimental law using the coefficient of restitution*
- *calculate the loss of mechanical energy in a collision*
- *recall and use the definition of impulse as a change of momentum*

LEARNING SUMMARY

Conservation of momentum

EDEXCEL · M2
OCR · M2
WJEC · M1

It's important to use the correct units. Remember that speed is the magnitude of the velocity.

Key points from AS

- **Momentum**
  Revise AS page 79

When particles coalesce, they stick together and behave as a single particle.

The velocity of B is in the negative direction.

The momentum of an object is the product of its mass and its velocity, i.e. momentum = $m\mathbf{v}$. Momentum has direction and so it is a vector quantity.

Momentum is measured in Ns (Newton seconds).

For example, an object of mass 5 kg moving with speed 6 ms^{-1} in a given direction has momentum $5 \times 6 = 30$ Ns in the direction of movement.

The **Principle of conservation of momentum** states that, in a collision, the total momentum before impact = the total momentum after impact.
Remember:

- always draw a diagram to represent the situation
- choose one direction to be positive (usually left to right $\rightarrow$)
- show any unknown velocities in the positive direction.

Example

Two particles A and B of masses 4 kg and 6 kg respectively move towards each other along the same straight line. Particle A is moving with speed 3 ms^{-1} and particle B is moving with speed 5 ms^{-1}. Given that the particles coalesce on impact, find their common velocity immediately afterwards.

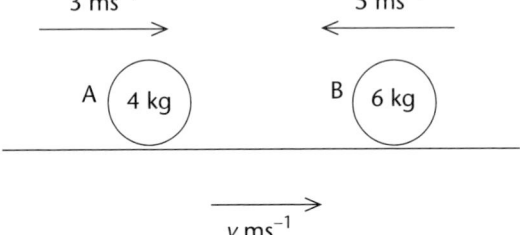

Before impact:

3 ms^{-1} A (4 kg)   5 ms^{-1} B (6 kg)

After impact: v ms^{-1}

By the principle of conservation of momentum:

$$4 \times 3 - 6 \times 5 = 10v$$

giving: $\qquad 10v = -18$

so: $\qquad v = -1.8$

v is negative so the direction of the velocity is from right to left.

The particles move with velocity 1.8 ms^{-1} in the direction from B to A.

Newton's experimental law

EDEXCEL M2
OCR M2
WJEC M1

Some problems require two unknown values to be found and in this situation a second equation is needed.

Newton found by experiment that, when two objects collide, the ratio of their speed of separation to their speed of approach has a fixed value. The symbol used to represent this ratio is e and its value is depends on what the objects are made of.

$$e = \frac{\text{speed of separation}}{\text{speed of approach}}$$

This is known as **Newton's experimental law** and is often used to provide a second equation.

This is often used in the form $e \times$ speed of approach = speed of separation.

The speed of separation is always less than or equal to the speed of approach so it follows that $0 \leqslant e \leqslant 1$. The ratio e is called the **coefficient of restitution**.

Example

In the exam, you may need to deal with more than one impact. See the sample exam questions p. 119.

A smooth sphere P moves towards a stationary smooth sphere Q in a straight line through their line of centres. P has mass 2 kg and speed 5 ms^{-1}. Q has mass 4 kg. Given that the coefficient of restitution between the spheres is 0.5, find:

(a) the speed of each sphere immediately after impact
(b) the loss of energy in the collision.

(a)

Before impact:

After impact:

5 ms^{-1}

P (2 kg)   Q (4 kg)

v_1 v_2

Momentum is conserved:
$2 \times 5 + 4 \times 0 = 2v_1 + 4v_2$
So:

$$v_1 + 2v_2 = 5 \qquad (1)$$

Newton's law:

$$0.5 \times 5 = v_2 - v_1 \qquad (2)$$

Adding (1) and (2) gives:
and:

$$3v_2 = 7.5 \implies v_2 = 2.5$$
$$v_1 = 0$$

Immediately after the impact, P is brought to rest and Q moves with speed 2.5 ms^{-1}.

Kinetic energy = $\frac{1}{2}mv^2$. Even though momentum is conserved, energy is lost.

(b) Loss of energy = KE before impact − KE after impact
$$= 0.5 \times 2 \times 5^2 - 0.5 \times 4 \times 2.5^2$$
$$= 12.5 \text{ J}$$

Impulse

EDEXCEL   M2
OCR   M2

Key points from AS

- **Impulse**
  Revise AS page 80

Impulse = change in momentum

When two particles collide, each receives an impulse from the other of equal size but opposite sign. In this way, the total change in momentum is zero. We would expect this to happen because the total momentum of the system is conserved in a collision.

Example

A ball of mass 0.5 kg moving with velocity $6\mathbf{i} + 2\mathbf{j}$ ms^{-1} is kicked and receives an impulse of $2\mathbf{i} + 11\mathbf{j}$ Ns. Find the velocity of the ball immediately after it is kicked.

Momentum immediately after kick = initial momentum + impulse

$$= 0.5(6\mathbf{i} + 2\mathbf{j}) + 2\mathbf{i} + 11\mathbf{j} = 5\mathbf{i} + 12\mathbf{j}$$
$$= 0.5\mathbf{v} \text{ (where } \mathbf{v} \text{ is the velocity)}$$

This gives: $\mathbf{v} = 10\mathbf{i} + 24\mathbf{j}$ ms^{-1}

Progress check

Two particles A and B move towards each other with speeds of 6 ms^{-1} and 2 ms^{-1} respectively. A has mass 3 kg and B has mass 1 kg. The coefficient of restitution between the particles is 0.6. Find the speed of each particle after the collision.

A 2.8 ms^{-1}, B 7.6 ms^{-1}.

3.7 Uniform circular motion

After studying this section you should be able to:

- *understand angular speed and be able to apply the formula $v = r\omega$*
- *understand that a particle moving in a circular path with constant speed has an acceleration towards the centre of the circle*
- *use the formula $\omega^2 r$, or its equivalent form $\dfrac{v^2}{r}$, to represent the magnitude of the acceleration towards the centre of the circle in solving problems*

LEARNING SUMMARY

Angular speed

AQA	M2
OCR	M2
WJEC	M2
CCEA	M2

Suppose that a particle P moves with constant speed in a circular path with centre O.

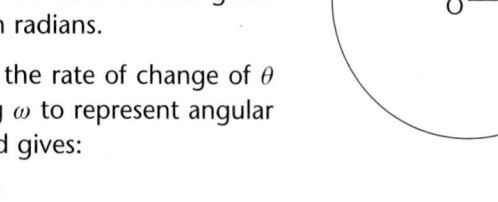

The direction of OP changes with time. The angle θ, through which OP turns in some given time, is usually measured in radians.

The **angular speed** of P is the rate of change of θ with respect to time. Using ω to represent angular speed in radians per second gives:

$$\omega = \frac{\theta}{t}$$

One advantage of using radians is that the relationship between angular speed ω, and linear speed v, is easily expressed as $v = rw$.

For example, a particle moving in a circular path of radius 2 m with angular speed 10 radians/sec has a straight line speed of $2 \times 10 = 20$ ms^{-1}.

Radial acceleration

AQA	M2
OCR	M2
WJEC	M2
ICCEA	M2

Notice that while the speed of the particle may be constant, its *direction* changes as it moves around the circle. This means that the *velocity* is *not constant* and so the particle must be *accelerating*.

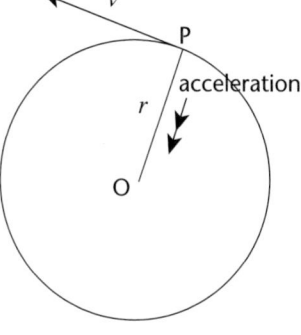

This acceleration is always directed towards the centre of the circle and is known as **radial acceleration** or sometimes **centripetal acceleration**.

The magnitude of this acceleration is given by $a = \omega^2 r$.

Since $v = rw$, it follows that $a = \dfrac{v^2}{r}$.

This means that there is a choice of which formula to use. In practice, you should use the one that best fits the available information.

Example

> Since we are given the angular speed, the formula $a = \omega^2 r$ is the simpler one to use here.

A particle moves in a circular path of radius 3 m with angular speed 5 radians/sec. Find the magnitude of its radial acceleration.

Using $a = \omega^2 r$:   the radial acceleration is $5^2 \times 3 = 75$ ms^{-2}.

Circular motion and force

AQA ▶ M2
OCR ▶ M2
WJEC ▶ M3
CCEA ▶ M2

The radial acceleration of an object moving in a circular path must be produced as the result of a *force acting towards the centre of the circle*. This force is sometimes called the **centripetal force**.

Using $F = ma$, the force needed to keep an object of mass m kg moving in a circular path is $m\omega^2 r$. An alternative form of this is $\dfrac{mv^2}{r}$.

In applying this theory to solving problems, the centripetal force may take various forms such as a tension in a string, friction or the gravitational attraction of a planet.

Example

One end of a string of length 80 cm is attached to a point O on a smooth horizontal table. The other end is attached to a mass of 3 kg moving with speed 4 ms⁻¹ and the string is taut. Find the tension in the string.

> It is important to write the length of the string in metres.

Using $T = \dfrac{mv^2}{r}$ gives $T = \dfrac{3 \times 4^2}{0.8} = 60$

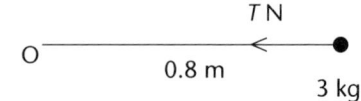

The tension in the string is 60 N.

The conical pendulum

AQA ▶ M2
OCR ▶ M2
WJEC ▶ M2
CCEA ▶ M2

One particular situation involving circular motion is described as the **conical pendulum**.

This is where an object moves in a horizontal circle while suspended by a string attached to a point vertically above the centre of the circle.

* The vertical component of the tension in the string supports the weight of the object.
* The horizontal component of the tension in the string provides the centripetal force.

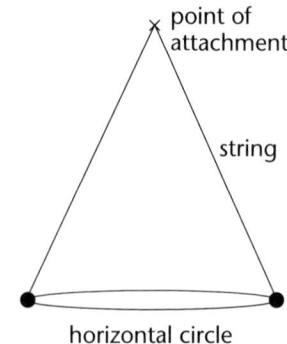

Example

A light inextensible string is fixed at one end to a point O. The other end is attached to a particle of mass 2 kg which moves in a horizontal circle of radius 30 cm with angular speed 4 radians per second. Find the angle of inclination of the string to the vertical. Take $g = 9.8$ ms⁻².

Vertically: $T \cos \theta = 19.6$ (1)

Horizontally: $T \sin \theta = 2 \times 4^2 \times 0.3 = 9.6$ (2)

(2) ÷ (1) gives: $\tan \theta = \dfrac{9.6}{19.6} \Rightarrow \theta = 26.1°$

The string is inclined at 26.1° to the vertical.

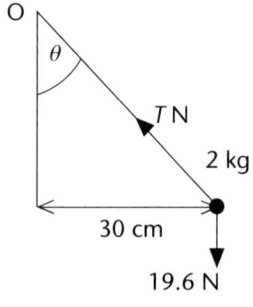

Progress check

One end of a string of length 60 cm is attached to a point O on a smooth horizontal table. The other end is attached to a mass of 5 kg moving with speed 3 ms⁻¹ and the string is taut. Find the tension in the string.

1. 75 N

3.8 Further circular motion

After studying this section you should be able to:

- *solve problems involving uniform motion in a horizontal circle on banked tracks*
- *solve problems involving motion in a vertical circle*

LEARNING SUMMARY

Banked tracks

AQA ▸ M3
OCR ▸ M2
WJEC ▸ M2
CCEA ▸ M2

When a car travels in a horizontal circle, on a flat surface, the centripetal force required to keep it moving in a circle is supplied by friction. The speed of the car must be restricted because the frictional force has a limiting value. Once this limiting value is reached, any further increase in speed will cause the car to skid.

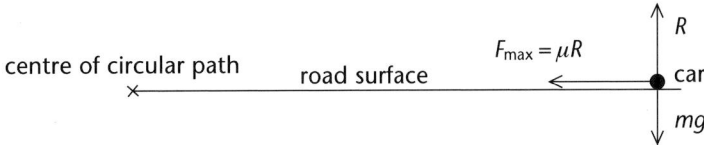

One way to increase the maximum speed of a car in a horizontal circle is to bank the track so that the reaction force of the road surface on the car has a horizontal component. This can result in a greater value of the maximum centripetal force which allows greater speeds to be reached without skidding.

Example

A car travels around a track in a horizontal circle of radius 150 m.

(a) What angle does the track need to be banked at so that the car has no tendency to side-slip at a speed of 25 ms^{-1}?

(b) Given that $\mu = 0.7$ find the greatest speed that the car can travel around the track without slipping.

(a) There is no frictional force, in this case, since there is no tendency to slip

> The forces balance vertically.

Vertically: $\qquad R \cos \theta = mg \qquad\qquad$ (1)

> Using $F = ma$ where the acceleration is towards the centre of the circle.

Horizontally: $\qquad R \sin \theta = \dfrac{m \times 25^2}{150} \qquad$ (2)

(2) ÷ (1) gives: $\qquad \tan \theta = \dfrac{25^2}{150 \times 9.8}$

$$\Rightarrow \theta = 23.0^0$$

The track needs to be banked at 23.0^0.

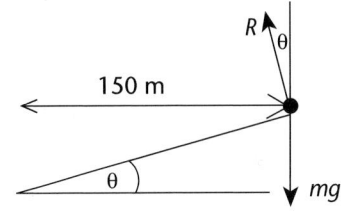

(b) Vertically: $R \cos 23^0 - 0.7R \sin 23^0 = mg$ (3)

Horizontally: $R \sin 23^0 + 0.7 R \cos 23^0 = \dfrac{mv^2}{150}$ (4)

(4) ÷ (3) gives: $\dfrac{v^2}{150g} = \dfrac{\sin 23° + 0.7 \times \cos 23°}{\cos 23° - 0.7 \times \sin 23°}$

$$= 1.5997 \ldots$$

so: $v^2 = 1.5997 \ldots \times 150 \times 9.8$

$$= 2351.76 \ldots \Rightarrow v = 48.49 \ldots$$

The maximum speed is 48.5 ms^{-1} to 3 s.f.

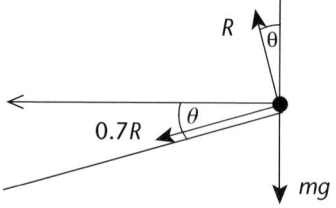

Motion in a vertical circle

AQA   M2
WJEC   M2

Solving problems regarding the motion of a particle in a vertical circle generally involves:

- calculation of the speed of the particle by considering energy
- consideration of the radial forces necessary to maintain circular motion.

> **KEY POINT.**
>
> For a particle attached to a string it is assumed that the only forces acting on the particle are its weight and the tension in the string. Since the tension always acts in a direction perpendicular to the motion *it does no work* and so the total mechanical energy of the particle is conserved.

Example

A particle P of mass 0.3 kg is attached to one end of a light inelastic string of length 0.8 m. The other end of the string is attached to a fixed point O. The particle is given an initial speed of 5 ms^{-1} horizontally from a point 0.8 m vertically below O. Find:

(a) the speed of P when OP is horizontal
(b) the tension in the string at this point
(c) the angle that OP makes with the upward vertical when the string first becomes slack.

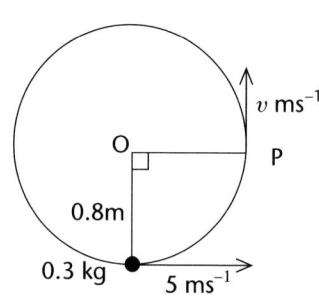

(a) Since energy is conserved:

$$\tfrac{1}{2} \times 0.3 \times v^2 = \tfrac{1}{2} \times 0.3 \times 5^2 - 0.3 \times 9.8 \times 0.8$$

giving: $v^2 = 5^2 - 2 \times 9.8 \times 0.8 = 9.32$

$$\Rightarrow v = 3.052 \ldots$$

$KE = \tfrac{1}{2} mv^2$
$GPE = mgh$

so the speed of P when OP is horizontal is 3.05 ms^{-1} to 3 s.f.

(b) Horizontally: $T = \dfrac{0.3 \times 9.32}{0.8} = 3.495$

The tension in the string is 3.50 N to 3 s.f.

(c) Using the conservation of energy principle:

$\frac{1}{2} \times 0.3 \times v^2 = \frac{1}{2} \times 0.3 \times 5^2 - 0.3 \times 9.8(0.8 + 0.8 \cos \theta)$

So $v^2 = 5^2 - 19.6 \times 0.8(1 + \cos \theta)$ (1)

When the string first becomes slack, the component of weight along PO is equal to the force needed to maintain circular motion. This gives:

$0.3 \times 9.8 \cos \theta = \dfrac{0.3v^2}{0.8} \Rightarrow v^2 = 0.8 \times 9.8 \cos \theta$

(2)

From (1) and (2) $\cos \theta = \dfrac{9.32}{23.52} \Rightarrow \theta = 66.7^0$

The string first becomes slack when OP makes an angle of 66.7^0 with the upward vertical.

Progress check

1   A car travels around a track in a horizontal circle of radius 200 m.
    What angle does the track need to be banked at so that the car has no
    tendency to side-slip at a speed of 30 ms^{-1}?

2   A particle P of mass 0.5 kg is attached to one end of a light inextensible string
    of length 0.6 m. The other end of the string is attached to a fixed point O. P is
    released from rest with OP horizontal and the string taut.
    Find the maximum tension in the string.

2   14.7 N
1   24.7^0

3.9 Work, energy and power

After studying this section you should be able to:

- *understand kinetic and potential energy and the work energy principle*
- *understand the principle of conservation of mechanical energy*
- *understand the definition of power*
- *solve problems involving work energy and power*

LEARNING SUMMARY

Work and energy

AQA	M2
EDEXCEL	M2
OCR	M2
WJEC	M2
CCEA	M2

> The situation where the force is not constant is dealt with in Mechanics 3 of this book.

In everyday language, **energy** is needed to do **work**. This idea is given a formal interpretation within mechanics.

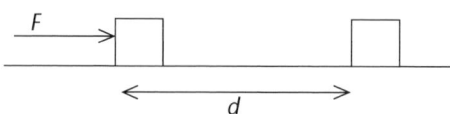

The work done, W, by a *constant* force, F, in moving an object through a distance, d, along its line of action is given by $W = Fd$. The unit of work is the **joule** (J).

Energy is also measured in joules and the energy used in moving the object is equal to the work done. In a way, work and energy are like two sides of the same coin.

Energy may take different forms such as heat, sound, electrical, chemical and mechanical. An important property of energy is that it may change from one form to another. For your mechanics module, the focus of attention is on mechanical energy.

There are two forms of mechanical energy: **kinetic energy (KE)** and **potential energy (PE)**

The kinetic energy of an object is the energy it has due to its *motion*.

Its value depends on the mass of the object and its speed. $KE = \frac{1}{2}mv^2$.

For example, a mass of 4 kg moving with speed 5 ms^{-1} has $KE = \frac{1}{2} \times 4 \times 5^2 = 50$ J

The potential energy of an object is the energy it has due to its *position*.

> Potential energy may be gravitational or elastic. Elastic potential energy is covered in Mechanics 3 in this book.

The **gravitational potential energy** of an object, relative to some level, is equal to the work done against gravity in raising the object from that level to its current position.

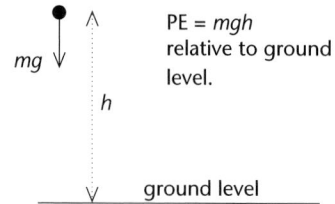

$PE = mgh$ relative to ground level.

> The total mechanical energy of an object is the sum of its KE and PE.
>
> The **principle of conservation of mechanical energy** states that the total mechanical energy of a system remains constant, provided that it is not acted on by any external force, other than gravity.

KEY POINT

The principle of conservation of energy applies, for example, to the motion of a projectile. Any change in the PE of the projectile corresponds to an equal but opposite change in its KE.

Power

AQA — M2
EDEXCEL — M2
OCR — M2
WJEC — M2
CCEA — M2

Power is the rate of doing work. It is measured in watts (W) where $1\text{ W} = 1\text{ Js}^{-1}$.

If the point of application of a force F moves with speed v in the *direction of the force* then the power P is given by $P = Fv$.

Example

Find the power generated by a car engine when the car travels at a constant speed of 72 kmh^{-1} on horizontal ground against resistance forces of 2500 N.

> Since the speed is constant, the force produced by the engine must match the resistance force.

$$72 \text{ kmh}^{-1} = \frac{72 \times 1000}{60 \times 60} \text{ ms}^{-1} = 20 \text{ ms}^{-1}.$$

$$\text{Power} = 2500 \times 20 \text{ W} = 50\,000 \text{ W}$$
$$= 50 \text{ kW}$$

Exam questions may involve resisted motion on an inclined plane.

Example

A car of mass 1000 kg travels up a hill inclined at angle θ to the horizontal, where $\sin\theta = \frac{1}{20}$. The non-gravitational resistance to motion is 2000 N and the power output from the engine is 60 kW. Find the acceleration of the car when it is travelling at 10 ms^{-1}. Take $g = 10$ ms^{-2}.

> A clearly labelled diagram is a must.

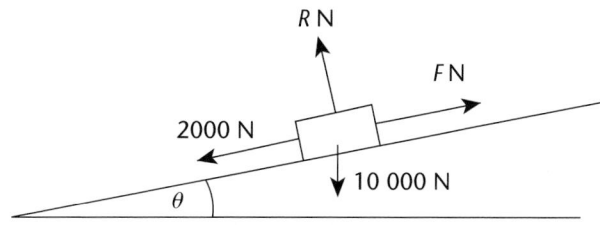

> In this formula, F represents the *resultant force* up the hill.

Using $P = Fv$: $60\,000 = F \times 10 \Rightarrow F = 6000$

Using $F = ma$: $6000 - 2000 - 10\,000 \times \frac{1}{20} = 1000a$

This gives: $3500 = 1000a \Rightarrow a = 3.5$

The acceleration of the car up the hill is 3.5 ms^{-2}.

Progress check

1 A ball is kicked from ground level with speed 15 ms^{-1} and just clears a wall of height 2 m. The speed of the ball as it passes over the wall is v ms^{-1}.
   (a) Assume that the ball has mass m kg and use the conservation of mechanical energy principle to write an equation.
   (b) Solve the equation to find v.

2 A car of mass 800 kg travels up a hill inclined at angle θ to the horizontal, where $\sin\theta = \frac{1}{20}$. The non-gravitational resistance to motion is 1600 N and the power output from the engine is 45 kW. Find the acceleration of the car when it is travelling at 15 ms^{-1}. Take $g = 10$ ms^{-2}.

2 1.25 ms^{-2}
(b) 13.6 ms^{-1}
1 (a) $\frac{1}{2}m \times 15^2 = m \times 9.8 \times 2 + \frac{1}{2}mv^2$

3.10 Elastic springs and strings

After studying this section you should be able to:

- *understand and apply Hooke's law, including the term modulus of elasticity*
- *find the energy stored in an elastic string or spring*
- *solve problems involving elastic strings and springs using the work-energy principle*

LEARNING SUMMARY

Hooke's law

AQA ▶ M2
WJEC ▶ M2

An elastic string will vary in length depending on its tension. Its **natural length** l, is its length when the tension is zero. The *extra* length of an elastic string due to a tension T within it is called the **extension** x.

Hooke's law states that the tension in an elastic string is proportional to the extension. This may be expressed as $T \propto x$ or

$$T = kx \qquad (1)$$

for some constant value k.

> When a string has doubled its length, the extension must equal the natural length.

The **modulus of elasticity** λ, is the **tension** required to make an elastic string *double its length*. It follows from (1) that $\lambda = kl$ and $k = \dfrac{\lambda}{l}$.

> Hooke's law is normally used in this form.

Substituting for k in (1): Hooke's law may now be written as $T = \dfrac{\lambda}{l} x$.

Hooke's law applies in the same way to **springs** that are under tension. The difference is that a spring will also exert an outward force, or thrust, when it is *compressed* by a distance x. The formula given by Hooke's law applies in both situations.

Example

A light elastic string of natural length 0.8 m and modulus of elasticity 30 N is extended to a length of 1.1 m. Find the tension in the string.

In this case: $\quad l = 0.8$, $\lambda = 30$ and $x = 0.3$.

Using Hooke's law gives: $\qquad T = \dfrac{30}{0.8} \times 0.3 = 11.25$

The tension in the string is 11.25 N.

Example

A spring of natural length 0.5 m is compressed to a length of 0.35 m by a force of 60 N. Find the modulus of elasticity.

For the spring: $\quad l = 0.5$, $x = 0.15$ and $T = 60$.

Using Hooke's law gives: $\quad 60 = \dfrac{\lambda}{0.5} \times 0.15 \implies 0.3\lambda = 60$

so: $\qquad\qquad\qquad\qquad\qquad \lambda = 200$

The modulus of elasticity for the spring is 200 N.

Elastic potential energy

AQA ▸ M2
WJEC ▸ M2

The elastic potential energy EPE of an elastic string or spring is its capacity for doing work as it returns to its natural length. Assuming that there is no loss of mechanical energy, this is equivalent to the work done in producing the extension or compression in the first place.

> The area under the graph represents the work done against the tension in producing a given extension.

The diagram shows the linear relationship between T and x.

Work done $= \dfrac{1}{2} \times x \times \dfrac{\lambda}{l} x$

so the EPE $= \dfrac{\lambda x^2}{2l}$

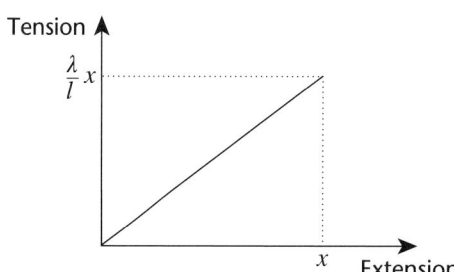

Example
An elastic string of natural length 1 m and modulus of elasticity 30 N is stretched to a length of 1.4 m. Calculate the EPE stored in the string.

> EPE is measured in Joules.

$$\text{EPE} = \frac{\lambda x^2}{2l} = \frac{30 \times 0.4^2}{2 \times 1} = 2.4 \text{ J.}$$

Example
One end of a light elastic string is attached to a point O, on a smooth horizontal table. The other end is attached to a particle P of mass 0.3 kg held in position on the table such that OP = 1.8 m. The string has modulus of elasticity 20 N and natural length 1.2 m.

The system is released from rest. Find the speed of P when the string becomes slack.

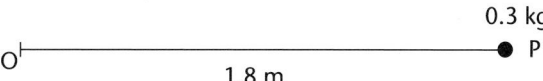

$$\text{EPE} = \frac{\lambda x^2}{2l} = \frac{20 \times 0.6^2}{2 \times 1.2} = 3 \text{ J.}$$

Since the table is *smooth*, all of the EPE will be converted to KE as the string returns to its natural length and becomes slack.

$\text{KE} = \frac{1}{2} mv^2 = \frac{1}{2} \times 0.3 \times v^2 = 0.15v^2$.

This gives: $0.15v^2 = 3 \implies v = \sqrt{20}$

P has speed 4.47 ms^{-1} when the string becomes slack.

Progress check

1 An elastic string of natural length 0.8 m is extended to a length of 1.4 m by a force of 60 N. Find the modulus of elasticity of the string.
2 A light spring with modulus of elasticity 80 N and natural length 0.9 m is compressed by a distance of 0.2 m. One end of the spring is fixed and a particle of mass 0.1 kg is attached to the other end. The system is released from rest on a smooth horizontal surface.
   Find the maximum speed reached by the particle in the subsequent motion.

2 5.96 ms^{-1} to 3 s.f.
1 80 N

Sample questions and model answers

1

A ball of mass 250 g is released from rest 2 m above the ground. The coefficient of restitution between the ball and the ground is 0.7. Take $g = 9.8 \text{ ms}^{-2}$.

(a) Find the height reached by the ball as it rebounds from the ground.

(b) Find the impulse exerted on the ball by the ground.

(a) Using $\quad v^2 = u^2 + 2as$

gives $\quad v^2 = 2 \times 9.8 \times 2$

$\qquad = 39.2$

so $\quad v = \sqrt{39.2}$

> There is no need to work out $\sqrt{39.2}$ at this stage.

The ball strikes the floor with speed $\sqrt{39.2} \text{ ms}^{-1}$.

Using Newton's law, the ball rebounds with speed $\sqrt{39.2} \times 0.7 \text{ ms}^{-1}$

> Some simple statements help to make your method clear.

Using $\qquad v^2 = u^2 + 2as$

gives $\qquad 0 = 39.2 \times 0.49 - 2 \times 9.8 \times h$

so $h = \dfrac{39.2 \times 0.49}{2 \times 9.8} = 0.98$

The ball reaches a height of 0.98 m when it rebounds from the ground.

(b) Taking upwards as positive:

Momentum of ball immediately before it hits the ground

$\qquad = -0.25 \times \sqrt{39.2} \text{ Ns}$

> Write the mass in kg.

Momentum of ball immediately after it hits the ground

$\qquad = 0.25 \times \sqrt{39.2} \times 0.7 \text{ Ns}$

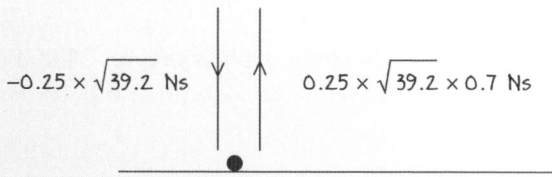

$-0.25 \times \sqrt{39.2} \text{ Ns} \qquad\qquad 0.25 \times \sqrt{39.2} \times 0.7 \text{ Ns}$

Impulse = change in momentum

$\qquad = 0.25 \times \sqrt{39.2} \times 0.7 - (-0.25 \times \sqrt{39.2})$

$\qquad = 2.66 \text{ Ns to 3 s.f.}$

The impulse exerted on the ball by the ground is 2.66 Ns.

Sample questions and model answers (continued)

2

A particle P of mass $3m$ is moving with speed $2u$ when it strikes a particle Q of mass m which is at rest. The coefficient of restitution between the particles is e.

(a) Show that the speed of P after the collision is $\frac{v}{2}(3-e)$ and find the speed of Q.

(b) Q subsequently strikes a wall perpendicular to its direction of motion and rebounds to so that a second collision with P occurs. Given that the coefficient of restitution between Q and the wall is $\frac{1}{3}$ find the speed of P after the second collision.

(c) Show that P moves in the same direction after the second collision.

(a) Before impact

$$\xrightarrow{\quad 2u \quad}$$

P $\left(3m\right)$ Q $\left(m\right)$

After impact

$$\xrightarrow{\quad v_P \quad} \qquad \xrightarrow{\quad v_Q \quad}$$

> The approach is exactly the same as when numerical values of the speeds are given.

Conservation of momentum gives: $6mu = 3mv_P + mv_Q$

which simplifies to $\qquad 6u = 3v_P + v_Q \qquad (1)$

Newton's experimental law gives: $2ue = v_Q - v_P \qquad (2)$

(1) − (2) gives: $\qquad 6u - 2ue = 4v_P$

> This is the result given in the question.

giving $\qquad v_P = \dfrac{u}{2}(3-e)$ as required

and from (2) $\qquad v_Q = 2ue + v_P = 2ue + \dfrac{u}{2}(3-e)$

$$= \frac{u}{2}(4e + 3 - e) = \frac{3u}{2}(e+1)$$

The speed of Q after the collision is $\dfrac{3u}{2}(e+1)$.

(b) The speed of Q after hitting the wall is $\dfrac{1}{3} \times \dfrac{3u}{2}(e+1) = \dfrac{u}{2}(e+1)$ in the opposite direction.

$$\xrightarrow{\quad \frac{u}{2}(3-e) \quad} \qquad \xleftarrow{\quad \frac{u}{2}(e+1) \quad}$$

> Repeat the process with the new speeds. Draw a diagram and set your working out clearly.

Before impact

P $\left(3m\right)$ Q $\left(m\right)$

After impact

$$\xrightarrow{\quad w_P \quad} \qquad \xrightarrow{\quad w_Q \quad}$$

Conservation of momentum gives: $\dfrac{3mu}{2}(3-e) - \dfrac{mu}{2}(e+1) = 3mw_P + mw_Q$

which simplifies to $\qquad 4u - 2ue = 3w_P + w_Q \qquad (3)$

Newton's experimental law gives: $\left(\dfrac{u}{2}(3-e) + \dfrac{u}{2}(e+1)\right)e = w_Q - w_P$

which simplifies to $\qquad 2u = w_Q - w_P \qquad (4)$

(3) − (4) gives $\qquad 2u - 2ue = 4w_P$

> There is no need to find w_Q.

and $\qquad w_P = \dfrac{u}{2}(1-e)$

so the speed of P immediately after the second collision is $w_P = \dfrac{u}{2}(1-e)$.

(c) $1 - e \geqslant 0$ since $0 \leqslant e \leqslant 1$ so the direction of movement of P is not changed.

Practice examination questions

1 A ball is struck with speed 25 ms^{-1} from a point 50 cm above the ground. When it has travelled 20 m horizontally, the ball just clears a fence that is 3 m high.

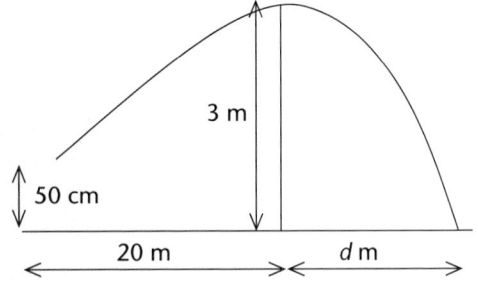

(a) Show that a possible angle of projection of the ball is 16.5°.

(b) Find the distance d m, between the fence and the point where the ball strikes the ground for this angle of projection.

2 A particle moves along the x axis such that at time t seconds its displacement from a point O on the line is $x = t^3 - 6t^2 + 3$ metres.

(a) Show that the particle is initially at rest and find its greatest distance from O in the negative direction.

(b) Find the speed of P when it passes through its initial position.

3 The position vector of a particle P at time t seconds relative to a fixed origin O is given by $\mathbf{r} = (4t^2 + 5)\mathbf{i} + 2t^3\mathbf{j}$ metres.

Find:

(a) the speed of P when $t = 1$

(b) the time when P is moving parallel to the vector $\mathbf{i} + 3\mathbf{j}$.

4 The diagram shows a uniform lamina ABCDEFGH.

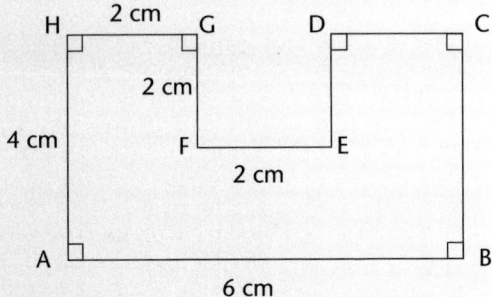

(a) Write down the distance of the centre of mass of the lamina from AH.

(b) Find the distance of the centre of mass of the lamina from AB.

(c) The lamina is now suspended freely from A. Find the angle that AB makes with the horizontal in its equilibrium position.

Practice examination questions (continued)

5

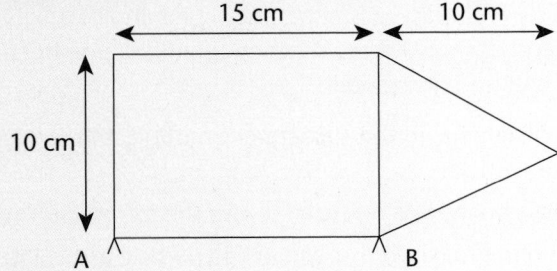

The diagram shows a uniform lamina of mass 0.8 kg resting on supports at A and B.

(a) Find the horizontal distance of the centre of mass from the support at A.

(b) Find the magnitude of the reaction force on the lamina at each support.

6

A and B are two smooth spheres of equal size and they are approaching each other on a direct line through their centres. A is moving with speed 4 ms^{-1} and B is moving with speed 2 ms^{-1}. The coefficient of restitution between the spheres is $\frac{1}{3}$.

Find:

(a) the speed of each sphere after the collision

(b) the impulse exerted by A on B during the collision

(c) the loss of mechanical energy during the collision.

7 Two particles P and Q approach each other with speeds u and $2u$ respectively. P has mass $3m$ and Q has mass m. The coefficient of restitution between the particles is e.

Given that the direction of motion of P is reversed in the collision, show that $e > \frac{1}{3}$.

8 A motorbike is driven round a corner in a circular path of radius 40 m. The maximum speed that this can be done without skidding is 15 ms^{-1}.

Calculate the coefficient of friction between the tyres and the road surface to 2 s.f.

Take $g = 9.8$ ms^{-2}.

9 One end of a light inelastic string of length 50 cm is attached to a fixed point 40 cm above a smooth horizontal table. The other end is attached to an object of mass 2 kg moving in a circular path of radius 30 cm on the table.

Given that the angular speed of the object is 3 radians per second, find:

(a) the magnitude of the acceleration of the object

(b) the tension in the string

(c) the reaction force of the table on the object.

10 A particle P of mass m is attached to one end of a light inelastic string of length l. The other end of the string is attached to a fixed point O. P is released from rest with the string taut and horizontal.

    (a) Explain why, in the subsequent motion, the tension in the string does no work on the particle.

    (b) Explain why it is not valid to use the formula $v^2 = u^2 + 2as$ in this situation.

    (c) Find the maximum speed of the particle assuming that there is no resistance to motion.

11 A car of mass 900 kg travels up a straight road inclined at angle α to the horizontal, where $\sin \alpha = \frac{1}{15}$. Assume that the total non-gravitational force opposing the car's motion has a constant value of 800 N. Take $g = 9.8$ ms^{-2}.

    (a) Find the driving force required for the car to maintain a steady speed up the slope.

    (b) Find the power output of the car's engine given that its speed up the slope has a constant value of 20 ms^{-1}.

    (c) At the top of the hill, the road surface is horizontal. If the power output of the engine remains the same, what is the initial acceleration of the car?

12 A football of mass 0.2 kg is travelling horizontally towards the goal with speed 20 ms^{-1} when it is struck by the keeper. It then rebounds vertically with speed 5 ms^{-1}.

Find the magnitude and direction of the impulse given to the ball by the keeper.

13 A particle P moves along a straight line passing through the point O. The displacement x metres, of the particle from O at time t seconds is given by:

$x = 4t^3 - 3t^2 + 7$ for $0 \leqslant t \leqslant 2$.

Find:

    (a) the velocity of the particle at time t

    (b) the distance that the particle travels before it changes direction

    (c) the time taken for the particle to return to its starting point.

14 A particle moving in a straight line is acted on by a single retarding force with magnitude proportional to the square of its speed. The initial speed of the particle is 20 ms^{-1} and its initial retardation is 3 ms^{-2}.

    (a) Show that the speed v ms^{-1} of the particle after t seconds is given by:

$$v = \frac{400}{3t + 20}$$

    (b) Given that the particle has mass 2 kg, find the magnitude of the work done by the retarding force between $t = 0$ and $t = 4$.

Practice examination questions (continued)

15 A light elastic string of natural length 2 m and modulus of elasticity $5mg$ has one end fixed at a point O. The other end is attached to an object of mass m which is released from rest at O.

Find:

(a) the distance that the object falls below O before being brought to instantaneous rest by the tension in the string

(b) the acceleration of the object at its lowest point.

16 A car is driven round a track in a horizontal circle of radius 80 m at a speed of 20 ms^{-1}. The track is banked at angle α to the horizontal and there is no tendency for the car to side-slip.

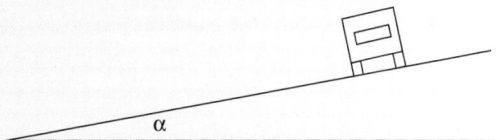

By representing the car as a particle:

(a) Find α to 1 d.p.

(b) The maximum speed that the car can travel on the banked track without slipping sideways is 28 ms^{-1}. Find the coefficient of friction between the tyres and the surface of the track.

17 The diagram shows a particle of mass 0.2 kg attached to one end of a light inelastic string of length 80 cm.

The particle moves in a horizontal circle on a smooth surface with the string inclined at 30° to the vertical.

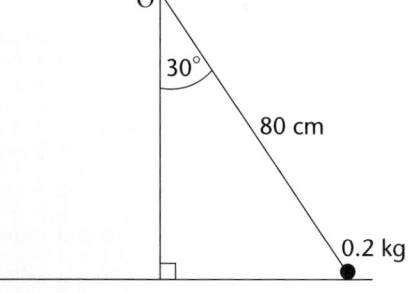

(a) Given that the speed of the particle is 1.2 ms^{-1} find the tension in the string and the reaction of the table on the particle.

(b) Find the minimum angular speed of the particle so that the reaction of the table on the particle is zero.

18 A small ball B is released from rest at the top of a horizontal circular tube. The ball travels down the outside of the tube's surface following a circular path, with centre C, until it falls away.

Given that the outside radius of the tube is 0.6 m and the mass of the ball is 0.1 kg find:

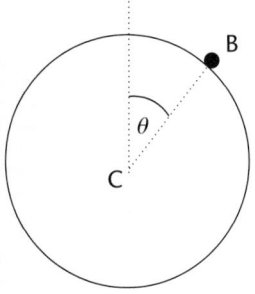

(a) an expression for the speed of B when CB makes an angle θ with the upward vertical

(b) the reaction of the tube on the ball in terms of θ and g

(c) the value of θ when the ball loses contact with the surface of the tube.
Take $g = 9.8$ ms^{-2}.

Chapter 4
Decision Mathematics 2

The following topics are covered in this chapter:

- Game theory
- Networks
- Critical path analysis

- Linear programming
- Matching and allocation

4.1 Game theory

After studying this section you should be able to:

- understand the idea of a zero-sum game and its representation using a pay-off matrix
- identify play-safe strategies and stable solutions
- find optimal mixed strategies for a game with no stable solution
- reduce a pay-off matrix using a dominance argument

LEARNING SUMMARY

A two person zero-sum game

EDEXCEL D2
OCR D2

The starting point in the mathematical theory of games is that the outcome of a game is determined by the **strategies** of the players.

A two person **zero-sum game** is a game in which the winnings of one player equal the losses of the other for every combination of strategies. Taking winnings to be positive and losses to be negative gives a zero sum in each case.

Viewing a game from one player's point of view, we could represent the outcomes (called pay-offs) for each combination of strategies in a matrix. This is called the **pay-off matrix** for that player.

Example

A and B are two players in a zero-sum game. A uses one of two strategies, W or X, and B uses one of the strategies Y or Z. The table shows the pay-off matrix for A.

The situation could equally be represented by the pay-off matrix for B. This would show corresponding values with opposite signs since this is a zero-sum game.

Pay-off matrix for A		B	
		Y	Z
A	W	2	−2
	X	5	−4

The pay-off matrix shows that if B adopts strategy Y then the pay-off for A will be 2 by using strategy W and 5 using strategy X. On the other hand, if B adopts strategy Z then the pay-off for A will be −2 using strategy W and −4 using strategy X.

The idea is that neither player knows in advance which strategy the other will use.

The play-safe strategy

AQA   D2
EDEXCEL   D2
OCR   D2

The **play-safe** strategy for a player is the strategy for which the minimum pay-off is as high as possible. In the example above, the minimum pay-off for A using strategy W is −2, whereas the minimum pay-off using strategy X is −4. This means that strategy W is the play-safe strategy for player A.

Notice that finding the play-safe strategy for player A involves comparing the minimum values in the *rows* of the pay-off matrix for A. Finding the play-safe strategy for B will involve comparing the values in the *columns*, remembering that B's pay-offs are the negatives of the ones in the pay-off matrix for A.

The minimum value for B using strategy Y is −5 and the minimum value using strategy Z is 2. This means that the play-safe strategy for B is strategy Z.

The situation is shown in the pay-off matrices for A and B.

Pay-off matrix for A	B Y	Z
A W	2	−2
A X	5	−4

Pay-off matrix for B	B Y	Z
A W	−2	2
A X	−5	4

In this case, the maximum of the minimum pay-offs, for each player, is in the corresponding position in the two matrices. This represents the **stable solution** to the problem referred to as the **saddle point** (or **minimax point**).

The solution is stable in the sense that neither player can improve their pay-off by taking a different strategy, given that the other player doesn't change. In other words, while B uses strategy Z, the best strategy for A is W and while A uses strategy W, the best strategy for B is Z.

> In the example, the sum of these values is −2 + 2 = 0 and the solution is stable.

> **KEY POINT**
> If the sum of the two values used to determine the play-safe strategies is **not zero** then the values cannot correspond to the same cell position in the play-off matrices. This means that there is no saddle point and the game has **no stable solution**.

Example

The pay-off matrices for two players in a zero-sum game are given below. Show that there is no stable solution for the game.

Pay-off matrix for P	Q Y	Z
P W	5	−3
P X	−4	6

Pay-off matrix for Q	Q Y	Z
P W	−5	3
P X	4	−6

The play-safe strategies for P and Q are shown shaded.
The values used to determine the play-safe strategies are −3 and −5.
There is no stable solution since $(-3) + (-5) \neq 0$.

Optimal strategies for games that are not stable

AQA   D2
EDEXCEL   D2
OCR   D2

The repeated use of the same strategy over a series of games is called a **pure strategy**. This provides the best results for both players in a game which has a stable solution. In the case where no stable solution exists, a **mixed strategy** is used in which each of the strategies is employed with a given probability to find the optimal solution.

Returning to the previous example:

Pay-off matrix for P		Q	
		Y	Z
P	W	5	−3
	X	−4	6

Suppose that P chooses:

strategy W with probability p

and   strategy X with probability $(1-p)$.

Then if:

Q chooses strategy Y, the expected *gain* for P is given by $5p - 4(1 - p) = 9p - 4$

Q chooses strategy Z, the expected *gain* for P is $-3p + 6(1 - p) = -9p + 6$

> *The optimal value occurs when these expressions are equal.*

$$9p - 4 = -9p + 6 \Rightarrow 18p = 10$$
$$\Rightarrow p = \tfrac{5}{9}$$

> *You could equally use $-9p + 6$ to get the value of the game.*

The value of the game is given by $9p - 4 = 1$.

Suppose that Q chooses:

strategy Y with probability q

and   strategy Z with probability $(1 - q)$.

Then if:

P chooses strategy W, the expected *loss* for Q is given by $5q - 3(1 - q) = 8q - 3$

P chooses strategy X, the expected *loss* for Q is $-4q + 6(1 - q) = -10q + 6$

$$8q - 3 = -10q + 6 \Rightarrow 18q = 9$$
$$\Rightarrow q = \tfrac{1}{2}$$

(As a check, the value of the game is $8q - 3 = 1$ as before).

This shows that the optimal strategy for both players is to use mixed strategies such that:

P chooses strategy W with probability $\tfrac{5}{9}$ and strategy X with probability $\tfrac{4}{9}$

Q chooses strategy Y and strategy Z with equal likelihood.

The expected long-term gain then for P, as an average per game, is 1 and this is also the expected long-term loss, as an average per game, for Q.

Graphical representation

AQA   D2
EDEXCEL   D2
OCR   D2

Each graph corresponds to a strategy for Q (i.e. the opponent of P).

The diagram shows graphs of $9p - 4$ and $-9p + 6$ against values of p from 0 to 1.

The point of intersection corresponds to the probability that gives the optimal mixed strategy for P in the last example.

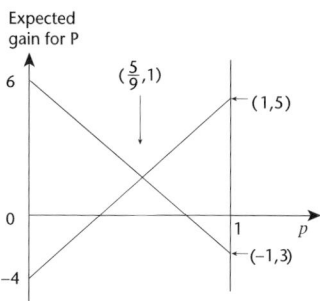

> **KEY POINT**
>
> When P's opponent has more strategies there will be more graphs with several points of intersection. You will need to identify the one that represents the optimal mixed strategy for P.
> *This will be the highest point on or below each of the graphs.*

This diagram represents a situation where P's opponent has three strategies to choose from.

The point representing the optimal mixed strategy for P is circled.

Notice how the problem of identifying the right vertex can be expressed as a linear programming problem in which the object is to maximise the expected gain for P subject to the constraints represented by the regions bounded by the straight line graphs.

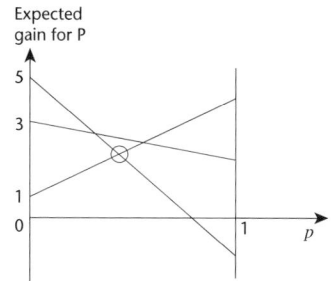

This is, in fact, the approach used for higher dimensional problems. The conditions are formulated as a linear programming problem which may then be solved by the simplex algorithm.

> **KEY POINT**
>
> If the pay-offs for one strategy are *always* better than the corresponding pay-offs for some other strategy then the weaker one can be ignored when determining the probabilities for a mixed strategy. In this way, the pay-off matrix is reduced by what is called a **dominance argument**.

Progress check

The table shows a pay-off matrix for player A in a zero-sum game.

(a) Find the play-safe strategy for each player.
(b) Show that this game has no stable solution.
(c) Find the best strategy for each player and the value of the game.

Pay-off matrix for A		Q	
		Y	Z
A	W	6	−4
	X	−3	5

(a) A(strategy X), Q(strategy Z)
(b) $(-3) + (-5) \neq 0$
(c) A Use W and X with probabilities $\frac{4}{9}$ and $\frac{5}{9}$
Q Use Y and Z with equal probability, value of game = 1

4.2 Networks

After studying this section you should be able to:

- *understand flows in networks and be able to find the maximum flow for a network involving multiple sources and sinks*
- *understand the travelling salesman problem and calculate upper and lower bounds for the total distance*

LEARNING SUMMARY

Flows in networks

AQA	D2
EDEXCEL	D1
OCR	D2

The **flows** referred to may be flows of liquids, gases or any other measurable quantities. The edges may represent such things as pipes, wires or roads that carry the quantities between the points identified as vertices.

A typical vertex has a flow into it and a flow out of it. The exceptions are a **source** vertex which has no input and a **sink** vertex which has no output.

Each edge of the network has a **capacity** which represents the maximum possible flow along that edge. In the usual notation, the capacity is written next to the edge and the flow is shown in a circle.

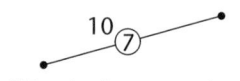

This edge has a capacity of 10 and a flow of 7.

The set of flows for a network is **feasible** if:

- The total output from all source vertices is equal to the total input for all sink vertices.
- The input for each vertex other than a source or sink vertex is equal to its output.
- The flow along each edge is less than or equal to its capacity.

A **cut** divides the vertices into two sets, one set containing the source and the other containing the sink.

The **capacity of a cut** is equal to the sum of the capacities of the edges that cross the cut, *taken in the direction from the source set to the sink set.*

In the diagram, the capacity of cut (i) is $15 + 10 = 25$.

The capacity of cut (ii) is $8 + 14 = 22$.

Notice that, for the second cut, the capacity of 4 is not included because it does not go from the source set to the sink set.

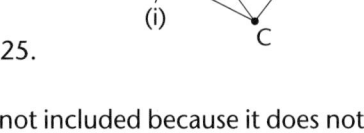

The maximum flow–minimum cut theorem

AQA	D2
EDEXCEL	D1
OCR	D2

The **maximum flow–minimum cut theorem** states that the maximum value of the flow through a network is equal to the capacity of the minimum cut. (This is a little bit like saying that a chain is only as strong as its weakest link.)

It follows from the theorem that if you find a flow that is equal to the capacity of some cut then it must be the maximum flow that can be established through the network.

For example, the capacity of cut (ii) for this network has already been found to be 22. The circled figures show one way of achieving this flow and the theorem now tells us that this must be the maximum possible flow through the network.

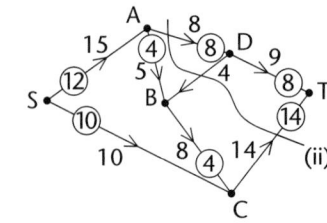

Flow augmentation

AQA   D2
EDEXCEL   D1
OCR   D2

Once an initial flow has been established through a network it is useful to know the extent to which the flow along any edge may be altered in either direction.

This is labelled as **excess capacity** and **back capacity** on each edge as shown. It may be possible to **augment** the flow through the network by making use of this flexibility along a path from the source vertex to the sink vertex. Such a path is called a **flow-augmenting path**.

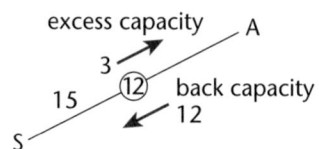

A **saturated** edge is one where the flow is equal to the capacity.

An algorithm for finding the maximum flow through a network by augmentation is:

Step 1   Find an initial flow by inspection.

Step 2   Label the excess capacity and back capacity for each edge.

Step 3   Search for a flow-augmenting path. If one can be found then increase the flow along the path by the maximum amount that remains feasible.

Step 4   Repeat steps 2 and 3 until no flow-augmenting path may be found.

Multiple sources and sinks

A network may have several sources S_1, S_2, ... and several sinks T_1, T_2, ... but the methods described in this section can still be applied through the use of a **supersource** and a **supersink**. A supersource S is an extra node, added to the network, that acts as a source for each of the original sources. Similarly, a supersink T is a node, added to the network, that acts as a sink for each of the original sinks.

Example

The network below shows two sources S_1 and S_2 and two sinks T_1 and T_2. Find the maximum flow through the network and state any flow-augmenting paths.

Explain why the flow is maximal.

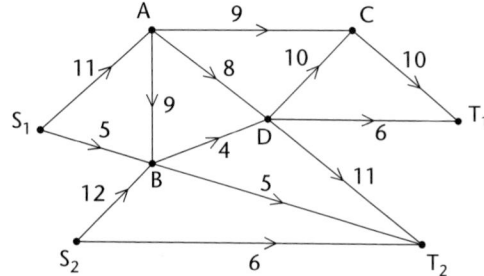

The first step is to add the supersource S and the supersink T to the network.

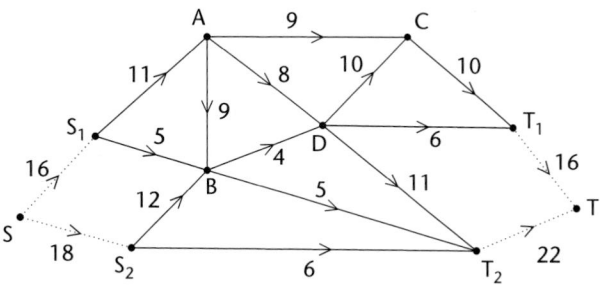

The usual convention is to use dotted lines for the edges linking S and T to the network.

The capacities on the edges SS_1 and SS_2 have been chosen to support the maximum possible flows from S_1 and S_2. Similarly, the capacities on the edges T_1T and T_2T have been chosen to support the maximum flows into T_1 and T_2. The diagram also shows a cut with capacity $11 + 4 + 5 + 6 = 26$.

A flow of 26 can be achieved as shown in the table.

Flow-augmenting path	Flow	Saturated edges
SS_1ACT_1T	9	S_1A, AC
SS_1ADT_1T	2	S_1A, DT_1
SS_2BDT_1T	4	BD, DT_1
SS_2BT_2T	5	BT_2
SS_2T_2T	6	S_2T_2
Total	26	

The flows on the flow-augmenting paths are shown circled on the network below. The flow of 26 is maximal since it equals the capacity of the cut (maximum flow minimum cut theorem).

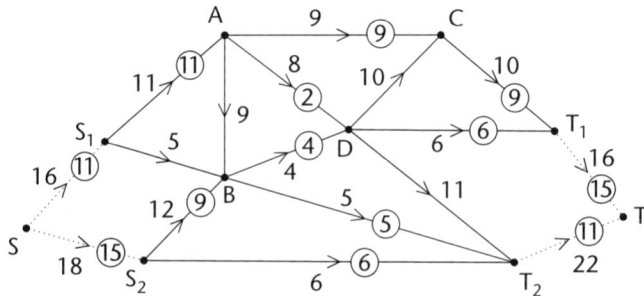

Note that each of the edges on the minimum cut is saturated. This will always be the case.

The final stage is to remove the supersource and supersink, along with the extra arcs, to show the maximum flow pattern for the original network.

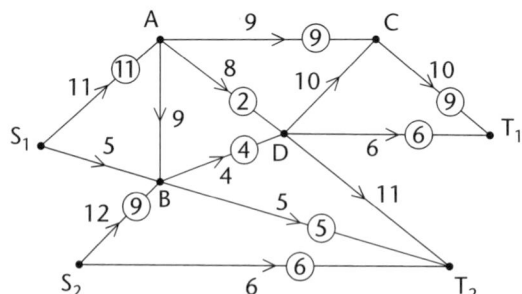

The travelling salesman problem

AQA ▶ D1
EDEXCEL ▶ D2
OCR ▶ D1

This is based on the classic situation of a salesman who wishes to visit a number of towns and return home using the shortest possible route.

The **travelling salesman problem** (TSP) is the problem of finding a route of minimum distance that visits every vertex and returns to the start vertex. For a small network it is possible to produce an exhaustive list of all possible routes and choose the one that minimises the total distance. For a large network an exhaustive check is not feasible, even with a computer, because the number of possible routes grows so rapidly.

It is useful to know within what limits the total distance must lie and there are algorithms that can be used for this purpose.

Finding an upper bound

> An upper bound for the total distance involved in the travelling salesman problem is given by twice the minimum spanning tree.
>
> **KEY POINT**

Key points from AS

- **Revise Prim's algorithm and Kruskal's algorithm** page 131

The minimum spanning tree (MST) for the network ABCD may be found using Prim's algorithm or Kruskal's algorithm.

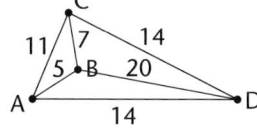

The total length of the MST is $5 + 7 + 14 = 26$.

An upper bound for the TSP is $2 \times 26 = 52$. This corresponds to the route ABCDCBA.

In some cases you might be able to find more than one short-cut that will enable you to reduce the upper bound.

However, an improved (smaller) upper bound can be found by using a **short-cut** from D directly back to A. The route is then ABCDA and the corresponding upper bound is 40.

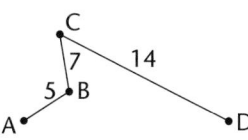

A different approach is to use the **nearest neighbour algorithm**.

Step 1   Choose a starting vertex.

Step 2   Move from your present position to the nearest vertex not yet visited.

Step 3   Repeat step 2 until every vertex has been visited. Return to the start vertex.

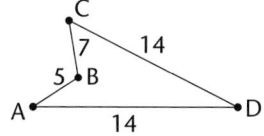

For the network ABCD considered above, the nearest neighbour algorithm gives the same result as using the MST with a short-cut. This won't always be the case.

Finding a lower bound

Step 1   Choose a vertex and delete it along with all of the edges connected to it.

Step 2   Find a minimum spanning tree for the remaining part of the network.

Step 3   Add the length of the MST to the lengths of the two shortest deleted edges.

The value found at step 3 is a lower bound for the travelling salesman problem. It may be possible to find a better (larger) lower bound by choosing to delete a different vertex initially and repeating the process.

Deleting D to start with gives a lower bound of 40. This corresponds to a cycle ABCDA and so represents the solution of the TSP.

Using the network ABCD again and deleting A gives the minimum spanning tree BCD of length $7 + 14 = 21$. Adding the two shortest deleted lengths gives a lower bound of 37.

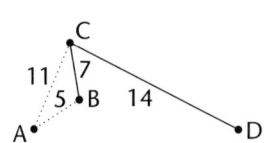

Progress check

1 Find the maximum flow through this network. Verify your answer using the maximum flow-minimum cut theorem.

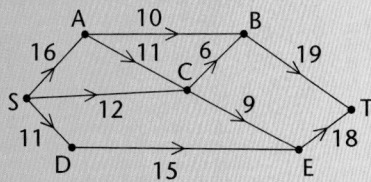

2 Use a supersource and supersink to find the maximum flow through this network.

Establish the flow through B in this case.

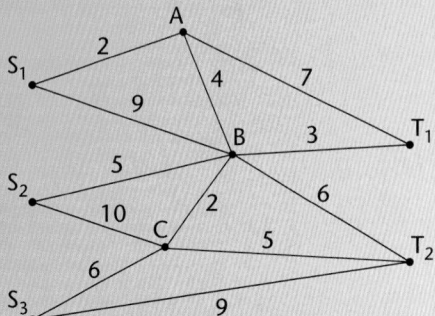

4.3 Critical path analysis

After studying this section you should be able to:

- construct an activity network from a given precedence table including the use of dummies where necessary
- use forward and backward scans to determine earliest and latest event times
- find the critical path in an activity network
- calculate the total float of an activity
- construct a chart for the purpose of scheduling

LEARNING SUMMARY

The process of representing a complex project by a network and using it to identify the most efficient way to manage its completion is called **critical path analysis**.

Activity networks

AQA   D2
EDEXCEL   D1
OCR   D2

A complex project may be divided into a number of smaller parts called **activities**. The completion of one or more activities is called an **event**.

Activities often rely on the completion of others before they can be started.

The relationship between these activities can be represented in a **precedence table**, sometimes called a **dependency table**.

In the precedence table shown on the right, the figures in brackets represent the **duration** of each activity, i.e. the time required, in hours, for its completion.

A precedence table can be used to produce an **activity network**. In the network, activities are represented by arcs and events are represented by vertices.

The vertices are numbered from 0 at the **start vertex** and finishing at the **terminal vertex**.

Activity	Depends on
A(3)	–
B(5)	–
C(2)	A
D(3)	A
E(3)	B, D
F(5)	C, E
G(1)	C
H(2)	F, G

The direction of the arrows shows the order in which the activities must be completed.

There must only be *one* activity between each pair of events in the network. The notation (i, j) is used to represent the activity between events i and j.

A **dummy activity** is one that has zero duration. A dummy is needed in this network to show that G depends on C whereas F depends on C *and* E.

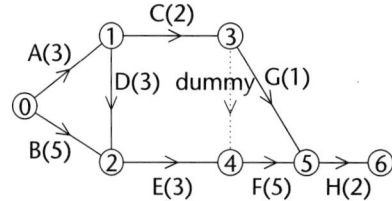

A dummy is shown with a dotted line. Its direction is important in defining dependency. In this case it shows that F depends on C not that G depends on E.

Earliest event times

The **earliest event time** for vertex i is denoted by e_i and represents the earliest time of arrival at event i with all dependent activities completed. These times are calculated using a **forward scan** from the start vertex to the terminal vertex.

Latest event times

The **latest event time** for vertex i is denoted by l_i and represents the latest time that event i may be left without extending the time for the project. These times are calculated using a **backward scan** from the terminal vertex back to the start vertex.

The **critical path** is the longest path through the network. The activities on this path are the **critical activities**. If any critical activity is delayed then this will increase the time needed to complete the project. The events on the critical path are the **critical events** and for each of these $e_i = l_i$.

It is useful to add the information about earliest and latest times to the network. The critical path is then easily identified.

Notice that B(5) lies between two critical events but it is not a critical activity.

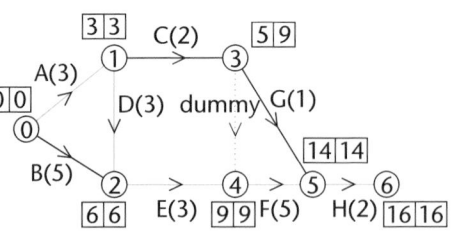

The **total float of an activity** is the maximum time that the activity may be delayed without affecting the length of the critical path. It is given by:

The total float of any critical activity is always zero.

latest finish time – earliest start time – duration of the activity.

Scheduling

AQA   D2
EDEXCEL   D1
OCR   D2

The process of allocating activities to workers for completion, within all of the constraints of the project, is known as **scheduling**.

The information regarding earliest and latest times for each activity is crucial when constructing a schedule. This information may be presented as a table or as a chart.

Typically, the purpose of scheduling is to determine the number of workers needed to complete the project in a given time, or to determine the minimum time required for a given number of workers to complete the project.

Activity	Duration	Start		Finish		Float
		Earliest	Latest	Earliest	Latest	
A(0, 1)	3	0	0	3	3	0
B(0, 2)	5	0	1	5	6	1
C(1, 3)	2	3	7	5	9	4
D(1, 2)	3	3	3	6	6	0
E(2, 4)	3	6	6	9	9	0
F(4, 5)	5	9	9	14	14	0
G(3, 5)	1	5	13	6	14	8
H(5, 6)	2	14	14	16	16	0

The critical activities are shown along one line.

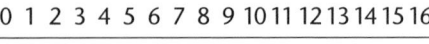

The diagram shown here is known as a cascade chart or Gantt chart.

The diagram illustrates the degree of flexibility in starting activities B, C and G. Remember that G cannot be started until C has been completed.

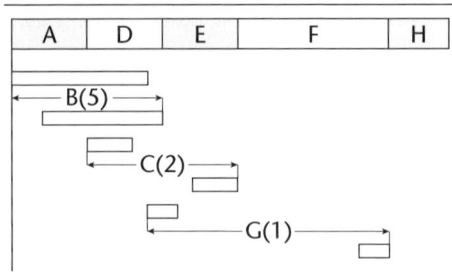

Progress check

Determine the critical activities and the length of the critical path for this network.

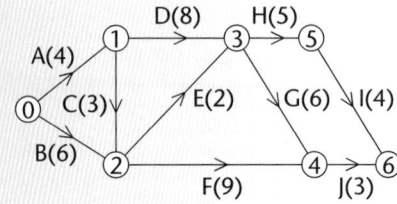

A, D, G and I. 21

4.4 Linear programming

After studying this section you should be able to:

- formulate a linear programming problem in terms of decision variables
- use a graphical method to represent the constraints and solve the problem
- use the Simplex algorithm to solve the problem algebraically

LEARNING SUMMARY

Formulating a linear programming problem

AQA D1
EDEXCEL D1
OCR D1

x and y are often used for the variables.

Typically, this may be to maximise a profit or minimise a loss.

To formulate a linear programming problem you need to:

- Identify the **variables** in the problem and give each one a label.
- Express the **constraints** of the problem in terms of the variables. You need to include non-negativity constraints such as $x \geqslant 0$, $y \geqslant 0$.
- Express the quantity to be optimised in terms of the variables. The expression produced is called the **objective function**.

Example

A small company produces two types of armchair. The cost of labour and materials for the two types is shown in the table.

	Labour	Materials
Standard	£30	£25
Deluxe	£40	£50

The total spent on labour must not be more than £1150 and the total spent on materials must not be more than £1250. The profit on a standard chair is £70 and the profit on a deluxe chair is £100. How many chairs of each type should be made to maximise the profit?

In this case, the variables are the number of chairs of each type that may be produced. Using x to represent the number of standard chairs and y to represent the number of deluxe chairs, the constraints may be written as:

It's a good idea to simplify the constraints where possible.

$$30x + 40y \leqslant 1150 \implies 3x + 4y \leqslant 115$$
$$25x + 50y \leqslant 1250 \implies x + 2y \leqslant 50$$

and $\quad x \geqslant 0, \quad y \geqslant 0.$

Using P to stand for the profit, the problem is to maximise $P = 70x + 100y$.

The graphical method of solution

AQA D1
EDEXCEL D1
OCR D1

Each constraint is represented by a region on the graph. It's a good idea to shade out the *unwanted* region for each one. The part that remains unshaded then defines the **feasible region** containing the points that satisfy all of the constraints.

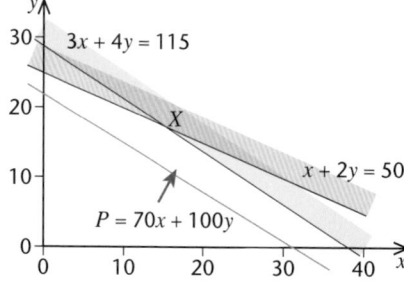

The blue line represents the points where the profit takes a particular value. Moving the line in the direction of the arrow corresponds to increasing the profit. This suggests that the maximum profit occurs at the point X.

X does not represent the solution in this case because both x and y must be integers.

Solving $3x + 4y = 115$ and $x + 2y = 50$ simultaneously gives X as $(15, 17.5)$.

The nearest points with integer coordinates in the feasible region are $(15,17)$ and $(14,18)$ The profit, given by $P = 70x + 100y$, is greater at $(14,18)$.

The maximum profit is made by producing 14 standard and 18 deluxe chairs.

The Simplex algorithm

AQA   D2
EDEXCEL   D1
OCR   D1

When there are three or more variables, a different approach is needed. The **Simplex algorithm** may be used to solve the problem algebraically. The information must be expressed in the right form before the algorithm can be used.

- First write the constraints, other than the non-negativity conditions, in the form $ax + by + cz \leqslant d$.
- Write the objective function in a form which is to be maximised.
- Add **slack variables** to convert the inequalities into equations.
- Write the information in a table called the **initial tableau**.

The objective function is already in a form which is to be maximised.

The constraints are in the right form.

Example

Find the maximum value of $P = 2x + 3y + 4z$ subject to the constraints:

$$3x + 2y + z \leqslant 10 \qquad \text{[i]}$$
$$2x + 5y + 3z \leqslant 15 \qquad \text{[ii]}$$
$$x \geqslant 0, \quad y \geqslant 0, z \geqslant 0.$$

Using slack variables s and t, the inequalities [i] and [ii] become:

$$3x + 2y + z + s = 10$$
$$2x + 5y + 3z + t = 15.$$

The objective function must be rearranged so that all of the terms are on one side of the equation to give:

$$P - 2x - 3y - 4z = 0.$$

The information may now be put into the initial tableau.

	P	x	y	z	s	t	value
The top row shows the objective function	1	−2	−3	−4	0	0	0
This row shows the first constraint	0	3	2	1	1	0	10
This row shows the second constraint	0	2	5	3	0	1	15

The columns for P, s and t contain zeros in every row apart from one. In each case, the remaining row contains 1 and the value of the variable is given in the end column of that row. The value of every other variable is taken to be zero. So, at this stage, the tableau shows that $P = 0$, $s = 10$, $t = 15$ and x, y and z are all zero. This corresponds to the situation at the origin.

You need to know how to read values from the tableau.

Using the algorithm is equivalent to visiting each vertex of the feasible region until an optimum solution is found. This occurs when there are no negative values in the objective row.

When the present tableau is not optimal, a new tableau is formed as follows:

- The column containing the most negative value in the objective row becomes the **pivotal column**. In this case the pivotal column corresponds to the variable z.
- Now divide each entry in the *value* column by the corresponding entry in the pivotal column provided that the pivotal column entry is positive.
  For the example above this gives $\frac{10}{1} = 10$ and $\frac{15}{3} = 5$
  The smallest of these results relates to the bottom row which is now taken to be the **pivotal row**. The entry lying in both the pivotal column and the pivotal row then becomes the **pivot**. In this case, the pivot is 3.

P	x	y	z	s	t	value
1	−2	−3	−4	0	0	0
0	3	2	1	1	0	10
0	2	5	3	0	1	15	← pivotal row

↑
pivotal column

- Divide every value in the pivotal row by the pivot, this makes the value of the pivot 1. The convention is to use fraction notation rather than decimals.

P	x	y	z	s	t	value
1	−2	−3	−4	0	0	0
0	3	2	1	1	0	10
0	$\frac{2}{3}$	$\frac{5}{3}$	1	0	$\frac{1}{3}$	5	← pivotal row

- Now turn the other values in the pivotal column into zeros by adding or subtracting multiples of the pivotal row.

See paragraph on preceding page about reading values from the tableau.

P	x	y	z	s	t	value
1	$\frac{2}{3}$	$\frac{11}{3}$	0	0	$\frac{4}{3}$	20
0	$\frac{7}{3}$	$\frac{1}{3}$	0	1	$-\frac{1}{3}$	5
0	$\frac{2}{3}$	$\frac{5}{3}$	1	0	$\frac{1}{3}$	5	← pivotal row

It is often necessary to repeat the process in order to find the optimum tableau.

- If there are no negative values in the objective row then the **optimum tableau** has been found. Otherwise choose a new pivotal column and repeat the process.

Each time through the process is called an **iteration**.

In this case the optimum tableau has been found after just one iteration. It shows that the maximum value of P is 20 and that this occurs when

$x = 0$, $y = 0$, $z = 5$, $s = 5$ and $t = 0$.

4.5 Matching and allocation

After studying this section you should be able to:

- use bipartite graphs to model matchings
- understand the conditions for matchings to be maximal or complete
- apply the maximum matching algorithm

LEARNING SUMMARY

Matchings and graphs

AQA · D1
EDEXCEL · D1
OCR · D2

A **bipartite graph** is a graph in which the vertices are divided into two sets such that no pair of vertices in the same set is connected by an edge.

In this case, the two sets are {A, B, C} and {p, q, r, s}.

Some vertices in a bipartite graph may not be connected to another vertex.

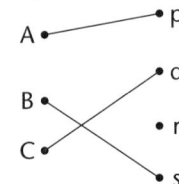

A **matching** between two sets may be represented by a bipartite graph in which there is at most one edge connecting a pair of vertices.

A **maximal matching** is a matching which has the maximum number of edges. This occurs when every vertex in one of the sets is connected to a vertex in the other set. The bipartite graph shown above represents a maximal matching.

A **complete matching** is a matching in which every vertex is connected to another vertex. This can only occur when the two sets contain the same number of vertices.

The matching improvement algorithm

AQA · D1
EDEXCEL · D1
OCR · D2

Figure 1 is a bipartite graph showing the possible connections between two sets. It does not represent a matching because some vertices have more than one connection.

Figure 2 is a bipartite graph representing an initial matching.

An initial matching may be improved by increasing the number of connections. This is the purpose of the **matching improvement algorithm**.

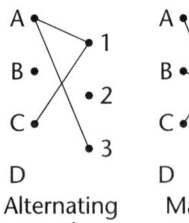

In an alternating path, the edges alternate between those from the bipartite graph that are not in the initial matching and those that are.

Step 1 In the initial matching, start from a vertex that is not connected and look for an **alternating path** to a vertex in the other set that is not connected.

Step 2 Each edge on the alternating path, not included in the initial matching, is now included and each edge originally included is removed.

Step 3 Repeat steps 1 and 2 using the latest matching in place of the initial matching until no further alternating paths can be found.

AQA · D2
EDEXCEL · D2
OCR · D2

Allocation problems

In an allocation (or assignment) problem, the object is to set up a matching that will optimise a particular objective, such as minimising a cost or maximising a profit. If we regard one set of objects as agents and the other set as tasks, then the underlying assumptions are that:

- Any agent may be assigned to any task (unlike the matching problems).
- Each agent is assigned at most one task and each task has at most one agent assigned to it.

When the number of agents is equal to the number of tasks, the problem is said to be balanced. If the number of agents exceeds the number of tasks, or vice versa, then the problem is unbalanced. The technique for solving an unbalanced problem involves introducing a dummy agent or task which then allows the solution to proceed as for the balanced case. In the solution of an unbalanced problem, a task assigned to a dummy agent isn't actually completed and an agent assigned to a dummy task doesn't actually have a task to carry out.

The opportunity cost matrix

The table shows the time in hours for three workers A, B and C to complete three separate tasks.

		Task		
		1	2	3
Worker	A	9	11	7
	B	8	10	9
	C	7	9	8

The cost matrix shows only the figures in hours. In this case, the 'cost' is measured in time.

9	11	7
8	10	9
7	9	8

The opportunity cost matrix is found from the cost matrix by:

1. Subtracting the smallest value in each row from every other value in that row.
2. Subtracting the smallest value in each column from every other value in that column.

9	11	7
8	10	9
7	9	8

$\xrightarrow{1}$

2	4	0
0	2	1
0	2	1

$\downarrow 2$

2	2	0
0	0	1
0	0	1

The Hungarian algorithm

The Hungarian algorithm is used to find the allocation that optimises the given objective.

Step 1 Find the opportunity cost matrix.

Step 2 Test for optimality. If true then make the allocation and stop.

Step 3 Revise the opportunity cost matrix and repeat step 2.

The method used to test for optimality is to find the minimum number of horizontal and/or vertical lines needed to cover all of the zeros in the opportunity cost matrix. If this number = n where the matrix has n rows and n columns, then the matrix will give an optimal allocation.

In the example, the number of lines needed = 3.
The matrix has 3 rows and columns so it will give an optimal allocation.

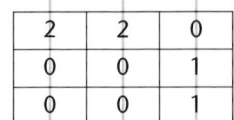

2	2	0
0	0	1
0	0	1

The optimal allocation is now found by selecting positions of zeros in the opportunity cost matrix such that one zero is selected in each row and in each column.

One possibility is shown here along with the corresponding figures in the original cost matrix.

2	2	**0**		9	11	**7**
0	**0**	1		8	**10**	9
0	0	1		**7**	9	8

The allocation is A → 3, B → 2 and C → 1 giving a total cost, in hours, of $7 + 10 + 7 = 24$.

In this case, there is an alternative solution.

2	2	**0**		9	11	**7**
0	0	1		**8**	10	9
0	**0**	1		7	**9**	8

The allocation is A → 3, B → 1 and C → 2 giving a total cost, in hours, of $7 + 8 + 9 = 24$.

Revising the opportunity cost matrix

When the opportunity cost matrix does not give an optimal solution, a revised opportunity cost matrix is found as follows:

Step 1 Find the smallest number not covered by a line. Subtract this number from every number not covered by a line.

Step 2 Add this number to every number lying at the intersection of two lines.

For example, this opportunity cost matrix has 3 rows and columns, but its zeros can be covered with just 2 lines.

0	3	1
0	2	5
0	0	0

Applying the steps given above produces this revised opportunity cost matrix.

0	2	**0**
0	1	4
1	**0**	0

Maximising allocation problems

If the objective involves maximising a quantity, such as a profit, then the first step is to subtract every value in the original matrix from a fixed value. This fixed value may be taken to be the largest value in the matrix or any value larger than this. The Hungarian algorithm is now applied in the usual way to this transformed matrix.

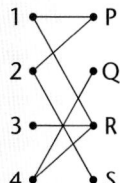

Sample questions and model answers

1

Ashley and Emma play a two-person zero-sum game in which they each consider three possible strategies. The table shows the pay-off matrix for Ashley.

		Emma		
		X	Y	Z
	A	5	−3	3
Ashley	B	4	1	−2
	C	−3	6	−4

(a) Find the play-safe strategy for each player and show that the game does not have a stable solution.

(b) Set up a linear programming problem to find Ashley's optimal mixed strategy. You are not required to solve it.

(a) The minimum row values are −3, −2 and −4 in Ashley's pay-off matrix. The maximum of these is −2 and so the pay-safe strategy for Ashley is strategy B.

The minimum column values in Emma's pay-off matrix are −5, −6 and −3. The maximum of these is −3 and so the play-safe strategy for Emma is strategy Z.

		Emma		
		X	Y	Z
	A	5	−3	3
Ashley	B	4	1	−2
	C	−3	6	−4

		Emma		
		X	Y	Z
	A	−5	3	−3
Ashley	B	−4	−1	2
	C	3	−6	4

The values use to determine the play-safe strategies do not have the corresponding positions in the pay-off matrices so there is no stable solution. Alternatively, $-3 + -2 \neq 0$ and so, again, there is no stable solution.

Sample questions and model answers (continued)

(b) Adding 5 to each of the figures in Ashley's pay-off matrix gives:

> The values need to be made positive by adding a fixed value to each one.

$$
\begin{array}{ccc}
10 & 2 & 8 \\
9 & 6 & 3 \\
2 & 11 & 1
\end{array}
$$

Using a, b and c to represent the probability that Ashley uses strategy A, B or C respectively, the linear programming problem is formulated as:

Maximise P subject to $a + b + c \leqslant 1$ where $a \geqslant 0$, $b \geqslant 0$, $c \geqslant 0$ and

$$P - 10a - 9b - 2c \leqslant 0$$
$$P - 2a - 6b - 11c \leqslant 0$$
$$P - 8a - 3b - c \leqslant 0$$

and P represents the value of the game.

2

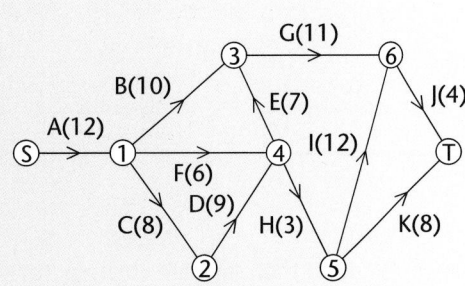

The diagram is an activity network for a project.

The figures in brackets represent the time in days to complete each activity.

The start vertex is S and the terminal vertex is T.

(a) Label each vertex with the earliest and latest time for the event. Give the length of the critical path.

(b) State which events are critical and give the critical path.

> Use a forward scan to find the earliest event times.

> Use a backward scan to find the latest event times.

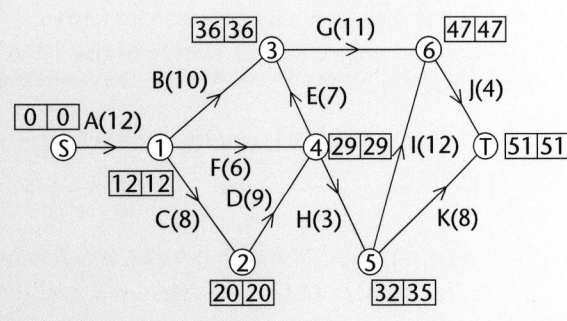

(a) The length of the critical path is 51 days.

(b) The critical events are S, 1, 2, 3, 4, 6 and T.
    The critical path is A, C, D, E, G and J.

> The length of the critical path is given by the earliest (and latest) event time at the terminal vertex.

Practice examination questions

1 A and B are the players in a two-person zero-sum game. Each may use one of two strategies when a game is played. The pay-off matrix for A is shown below.

		Player B	
		Y	Z
Player A	W	4	8
	X	6	3

(a) Find the best mixed strategy for each player.

(b) Find the value of the game.

2 Richard and Judy play a zero-sum game. The pay-off matrix for Richard is shown in the table.

		Judy		
		X	Y	Z
Richard	A	4	3	−4
	B	−3	6	8
	C	2	1	−6

(a) Find the play-safe strategy for each player.

(b) Show that there is no saddle point for the game.

(c) Explain why strategy C will not be part of Richard's mixed strategy.

(d) Find Richard's optimal mixed strategy.

3 A distribution manager has the task of transporting pallets of frozen food between two storage centres. The maximum number of pallets is to be delivered within a daily budget of £2000. Three types of van are available and each van can only do the trip once per day.

The details of what the vans can carry and the daily costs are shown in the table.

At most 10 drivers may be used in one day.

Van type	No. of pallets	Cost/day
Class A	4	£120
Class B	9	£300
Class C	12	£400

Let a represent the number of Class A vans used, b the number of Class B vans used, c the number of Class C vans used and P the total number of pallets transported in a day.

(a) Write an expression for the objective function.

(b) Formulate the task as a linear programming problem.

(c) Describe any special condition that the solution must satisfy.

Practice examination questions (continued)

4 The table shows the names of five employees of a company and the days when they are each available to work a late shift. The days that have been highlighted represent a first attempt at matching the employees to the available days to make a rosta.

(a) Use a bipartite graph to represent the relationship between employees and days available.

(b) Use a second bipartite graph to represent an initial matching based on the days highlighted.

(c) Implement a matching improvement algorithm to obtain a complete matching. Indicate the alternating path used.

Name	Days available
Abbie	**Mon**, Tue, Fri
Ben	Mon, **Wed**, Thu
Colin	Mon, **Fri**
Dave	Fri, Wed
Emma	**Thu**, Fri

5 A linear programming problem is formulated as:

Maximise $P = x + 2y + 3z$,

subject to $2x + 6y + z \leqslant 10$

$x + 4y + 5z \leqslant 16$

and $x \geqslant 0, y \geqslant 0, z \geqslant 0$.

(a) Set up an initial simplex tableau to represent the problem.

(b) Use the simplex algorithm to solve the problem.

State the corresponding values of x, y and z.

(c) Explain how you know that the optimum solution has been found.

Practice examination answers

Core 3 and Core 4 (Pure Mathematics)

1 $\dfrac{1}{(1+x)^2} = (1+x)^{-2}$

$\qquad = 1 + (-2)x + \dfrac{(-2)(-3)}{2!}x^2 + \dots$

$\qquad = 1 - 2x + 3x^2 + \dots \quad |x| < 1$

$\sqrt{4+x} = \sqrt{4\left(1+\dfrac{x}{4}\right)}$

$\qquad = 4^{\frac{1}{2}}\left(1 + \dfrac{x}{4}\right)^{\frac{1}{2}}$

$\qquad = 2\left(1 + \dfrac{x}{4}\right)^{\frac{1}{2}}$

$\qquad = 2\left(1 + (\tfrac{1}{2})\left(\dfrac{x}{4}\right) + \dfrac{(\frac{1}{2})(-\frac{1}{2})}{2!}\left(\dfrac{x}{4}\right)^2 + \dots\right)$

$\qquad = 2 + \dfrac{x}{4} - \dfrac{x^2}{64} + \dots \quad |x| < 4$

$\therefore \ f(x) = 1 - 2x + 3x^2 + 2 + \dfrac{x}{4} - \dfrac{x^2}{64} + \dots$

$\qquad = 3 - 1\tfrac{3}{4}x + 2\tfrac{63}{64}x^2 + \dots \quad |x| < 1$

Ignoring terms in x^3 and higher, $f(x) \approx a + bx + cx^2$, with $a = 3$, $b = -1\tfrac{3}{4}$, $c = 2\tfrac{63}{64}$.

The restriction is $|x| < 1$.

2   (a) $y = \cos^2 x \sin x$

$\qquad \dfrac{dy}{dx} = \cos^2 x(\cos x) + \sin x(-2 \sin x \cos x)$

$\qquad\qquad = \cos^3 x - 2 \cos x \sin^2 x$

$\qquad\qquad = \cos^3 x - 2 \cos x(1 - \cos^2 x)$

$\qquad\qquad = \cos^3 x - 2 \cos x + 2 \cos^3 x$

$\qquad\qquad = 3 \cos^3 x - 2 \cos x$

(b) $u = \cos x \Rightarrow \dfrac{du}{dx} = -\sin x$

$\qquad \dfrac{dx}{du} = -\dfrac{1}{\sin x} \Rightarrow \sin x \dfrac{dx}{du} = -1$

$\qquad \displaystyle\int \cos^2 x \sin x\, dx$

$\qquad = \displaystyle\int \cos^2 x \sin x \dfrac{dx}{du}\, du$

$\qquad = \displaystyle\int (-u^2)\, du$

$\qquad = -\tfrac{1}{3}u^3 + c$

$\qquad = -\tfrac{1}{3}\cos^3 x + c$

(This can be also be done by direct recognition, see page 48.)

3 $\dfrac{dy}{dx} = \dfrac{2x(x+1) - 1(x^2 - 4)}{(x+1)^2}$

$\qquad = \dfrac{x^2 + 2x + 4}{(x+1)^2}$

When $x = 2$, $\dfrac{dy}{dx} = \dfrac{4}{3}$

$\therefore$ gradient of normal $= -\tfrac{3}{4}$

Equation of normal at (2, 0):

$\qquad y - 0 = -\tfrac{3}{4}(x - 2)$

$\qquad 4y = -3x + 6$

$\quad 3x + 4y - 6 = 0$

$\therefore \ a = 3$, $b = 4$ and $c = -6$.

4   (a) $g(x) = (x+1)(x-1) = x^2 - 1$.

Since $x^2 \geqslant 0$ the range of g is $x \geqslant -1$.

(b) $fg(x) = f(x^2 - 1)$

$\qquad\qquad = 5(x^2 - 1) - 7$

$\qquad\qquad = 5x^2 - 12$.

(c) $fg(x) = f(x)$

$\qquad \Rightarrow 5x^2 - 12 = 5x - 7$

$\qquad \Rightarrow 5x^2 - 5x - 5 = 0$

$\qquad \Rightarrow x^2 - x - 1 = 0$

$\qquad \Rightarrow x = \dfrac{1 \pm \sqrt{5}}{2}$.

5   (a) $R \sin(x + a) \equiv 15 \sin x + 8 \cos x$

$\qquad \Rightarrow R \cos a = 15$ and $R \sin a = 8$ (a is acute)

$\qquad \Rightarrow R = \sqrt{15^2 + 8^2} = 17$

$\qquad \tan a = \tfrac{8}{15} \Rightarrow a = 28.07°$

So $15 \sin x + 8 \cos x \equiv 17 \sin(x + 28.07°)$.

(b)   Max value is 17.

(c) $15 \sin x + 8 \cos x = 12$

$\qquad \Rightarrow 17 \sin(x + 28.07°) = 12$.

$\qquad \sin^{-1}(\tfrac{12}{17}) = 44.90°$, $180° - 44.90° = 135.10°$.

$\qquad x + 28.07 = 44.90° \Rightarrow x = 16.8°$.

$\qquad x + 28.07 = 135.10° \Rightarrow x = 107.0°$.

Core 3 and Core 4 (Pure Mathematics) *(continued)*

6 $\quad x\dfrac{dy}{dx} = y + yx \Rightarrow x\dfrac{dy}{dx} = y(1 + x)$

Separating the variables:

$\dfrac{1}{y}\dfrac{dy}{dx} = \dfrac{1 + x}{x}$

$\dfrac{1}{y}\dfrac{dy}{dx} = \dfrac{1}{x} + 1$

$\displaystyle\int \dfrac{1}{y}\,dy = \int \left(\dfrac{1}{x} + 1\right)dx$

$\therefore \quad \ln y = \ln x + x + c$

When $x = 2$, $y = 4$,

$\Rightarrow \quad \ln 4 = \ln 2 + 2 + c$

$\Rightarrow \quad\quad c = \ln 2 - 2 \quad (\ln 4 = 2 \ln 2)$

So $\quad \ln y = \ln x + x + \ln 2 - 2$

$\ln\left(\dfrac{y}{2x}\right) = x - 2$

$\dfrac{y}{2x} = e^{x-2}$

$y = 2xe^{x-2}$

7 (a) (i) $\quad y = (4x - 3)^9$

$\Rightarrow \dfrac{dy}{dx} = 9(4x - 3)^8 \times 4$

$\Rightarrow \dfrac{dy}{dx} = 36(4x - 3)^8.$

(ii) $\quad y = \ln(5 - 2x)$

$\dfrac{dy}{dx} = \dfrac{1}{5 - 2x} \times (-2)$

$= \dfrac{-2}{5 - 2x} \quad \left(\text{you can write this as } \dfrac{2}{2x - 5}\right).$

(b) (i) $\displaystyle\int (4x - 3)\,dx = \dfrac{1}{36}(4x - 3)^9 + C.$

(ii) $\displaystyle\int_3^4 \dfrac{1}{5 - 2x}\,dx = -\dfrac{1}{2}\int_3^4 \dfrac{-2}{5 - 2x}\,dx$

$= -\dfrac{1}{2}\,[\ln|5 - 2x|]_3^4$

$= -\dfrac{1}{2}\,(\ln|-3| - \ln|-1|)$

$= -\dfrac{1}{2}\ln 3.$

8 (a) $\displaystyle\int_0^1 xe^{-2x}\,dx = \left[x(-\tfrac{1}{2}\,e^{-2x})\right]_0^1 - \int_0^1 (-\tfrac{1}{2}\,e^{-2x})\,dx$

$= -\tfrac{1}{2}\,e^{-2} + \tfrac{1}{2}\displaystyle\int_0^1 e^{-2x}\,dx$

$= -\tfrac{1}{2}\,e^{-2} + \tfrac{1}{2}\left[(-\tfrac{1}{2}\,e^{-2x})\right]_0^1$

$= -\tfrac{1}{2}\,e^{-2} + \tfrac{1}{2}(-\tfrac{1}{2}\,e^{-2} - (-\tfrac{1}{2}))$

$= -\tfrac{3}{4}\,e^{-2} + \tfrac{1}{4}$

(b) (i) $\quad y = xe^{-x}$

$\dfrac{dy}{dx} = x(-e^{-x}) + e^{-x} = e^{-x}(1 - x)$

$\dfrac{dy}{dx} = 0$ when $x = 1$

When $x = 1$, $y = e^{-1} \Rightarrow A$ is point $(1, e^{-1})$.

(ii) $\quad V = \pi\displaystyle\int_0^1 y^2\,dx = \pi\int_0^1 x^2 e^{-2x}\,dx$

$= \pi\left(\left[x^2(-\tfrac{1}{2}\,e^{-2x})\right]_0^1 - \displaystyle\int_0^1 -\tfrac{1}{2}\,e^{-2x}(2x)\,dx\right)$

$= \pi\left(-\tfrac{1}{2}\,e^{-2} + \displaystyle\int_0^1 xe^{-2x}\,dx\right)$

$= \pi(-\tfrac{1}{2}\,e^{-2} - \tfrac{3}{4}\,e^{-2} + \tfrac{1}{4})$ (using part (a))

$= \tfrac{\pi}{4}\,(1 - 5e^{-2})$

9 $\quad \dfrac{dP}{dt} \propto P \Rightarrow \dfrac{dP}{dt} = kP$, where $k > 0$.

Separating the variables

$\displaystyle\int \dfrac{1}{P}\,dP = \int k\,dt$

$\Rightarrow \quad \ln P = kt + c$

$\Rightarrow \quad\quad P = e^{kt+c}$

$\Rightarrow \quad\quad P = Ae^{kt}$ (where $A = e^c$)

$t = 0$, $P = 500 \Rightarrow 500 = A$

$t = 10$, $P = 1000 \Rightarrow 1000 = 500e^{10k}$

$\therefore \quad\quad 2 = e^{10k}$

$\Rightarrow \quad 10k = \ln 2$

$\Rightarrow \quad\quad k = 0.1 \ln 2$

$\therefore \quad P = 500e^{(0.1 \ln 2)t}$

When $t = 20$,

$P = 500e^{0.1 \ln 2 \times 20} = 500 \times 4 = 2000$

So the population is 2000 when $t = 20$.

Core 3 and Core 4 (Pure Mathematics) *(continued)*

10 (a)
$$x^3 + xy + y^2 = 7$$

$$3x^2 + x\frac{dy}{dx} + y + 2y\frac{dy}{dx} = 0$$

At $(1, 2)$

$$3 + \frac{dy}{dx} + 2 + 4\frac{dy}{dx} = 0$$

$$\Rightarrow 5\frac{dy}{dx} + 5 = 0 \text{ i.e. } \frac{dy}{dx} = -1$$

At $(1, 2)$ gradient of normal = 1

Equation of normal is

$$y - 2 = 1(x - 1)$$

i.e. $y = x + 1$

(b) $x = t^2, \; y = t + \dfrac{1}{t}.$

$$\frac{dx}{dt} = 2t$$

$$\frac{dy}{dt} = 1 - \frac{1}{t^2} = \frac{t^2 - 1}{t^2}$$

$$\frac{dy}{dx} = \frac{dy}{dt} \times \frac{dt}{dx} = \frac{t^2 - 1}{2t^3}$$

$$\frac{dy}{dx} = 0 \text{ when } t^2 - 1 = 0 \Rightarrow t = \pm 1$$

When $t = 1$, coordinates are $(1, 2)$.

When $t = -1$, coordinates are $(1, -2)$.

The stationary points are at $(1, 2)$ and $(1, -2)$.

11 (a) $\sin 2x = 0$ when $2x = 0, \pi, 2\pi, \ldots$

i.e. when $x = 0, \tfrac{1}{2}\pi, \pi, \ldots$

$\therefore a = \tfrac{1}{2}\pi$

(b) $A = \displaystyle\int_0^{\frac{1}{2}\pi} y\,dx = \int_0^{\frac{1}{2}\pi} \sin 2x \, dx$

$$= -\tfrac{1}{2}\left[\cos 2x\right]_0^{\frac{1}{2}\pi}$$

$$= -\tfrac{1}{2}(-1 - 1)$$

$$= 1$$

Area of R is 1 square unit.

(c) $V = \pi \displaystyle\int_0^{\frac{1}{2}\pi} y^2 \, dx$

$$= \pi \int_0^{\frac{1}{2}\pi} \sin^2 2x \, dx$$

$$= \tfrac{1}{2}\pi \int_0^{\frac{1}{2}\pi} (1 - \cos 4x)\,dx$$

$$= \tfrac{1}{2}\pi\left[x - \tfrac{1}{4}\sin 4x\right]_0^{\frac{1}{2}\pi}$$

$$= \tfrac{1}{2}\pi(\tfrac{1}{2}\pi - 0 - 0)$$

$$= \tfrac{1}{4}\pi^2$$

The volume generated is $\tfrac{1}{4}\pi^2$ cubic units.

12 Referring to origin $(0, 0)$

$\mathbf{a} = 2\mathbf{i} + 2\mathbf{j}$ and $\mathbf{b} = 2\mathbf{i} + 3\mathbf{j} + \mathbf{k}$

Also $\mathbf{b} - \mathbf{a} = \mathbf{j} + \mathbf{k}$

A vector equation for AB is

$\mathbf{r} = \mathbf{a} + \mu(\mathbf{b} - \mathbf{a})$ where μ is a scalar parameter

$\therefore \; \mathbf{r} = 2\mathbf{i} + 2\mathbf{j} + \mu(\mathbf{j} + \mathbf{k})$

If lines intersect, there are unique values for μ and λ for which

$2\mathbf{i} + 2\mathbf{j} + \mu(\mathbf{j} + \mathbf{k}) = 2\mathbf{j} + \mathbf{k} + \lambda(\mathbf{i} + \mathbf{k})$

Equating coefficients of $\mathbf{i}$, $\mathbf{j}$ and $\mathbf{k}$, in turn gives:

$$2 = \lambda$$

$$2 + \mu = 2$$

$$\mu = 1 + \lambda$$

There are no values of μ and λ that satisfy all three equations, so the lines have no common point.

The angle between the lines is the angle between their direction vectors.

$|\mathbf{j} + \mathbf{k}| = \sqrt{2}$ and $|\mathbf{i} + \mathbf{k}| = \sqrt{2}$

$(\mathbf{j} + \mathbf{k}).(\mathbf{i} + \mathbf{k}) = 1$

Let the angle be θ, then

$$\cos\theta = \frac{1}{\sqrt{2}\sqrt{2}} = 0.5$$

$$\Rightarrow \qquad \theta = 60°$$

Statistics 2

1 If $\mu = 20$, $\bar{X} \sim N(20, \frac{16}{10})$,

i.e. $\bar{X} \sim N(20, 1.6)$.

Carrying out a two-tailed test at 5% level, reject H_0 if $z < -1.96$ where

$$z = \frac{17.2 - 20}{\sqrt{1.6}} = -2.213\ldots$$

Since $z < -1.96$, reject H_0.

There is evidence that μ is not 20.

2 (a) If X is number of people with the health problem, then $X \sim B(80, 0.01)$. n is large and p is small, so $X \sim Po(np)$ i.e. $X \sim Po(0.8)$ approximately.

$$P(X < 3) = e^{-0.8} + 0.8\,e^{-0.8} + \frac{0.8^2}{2!}\,e^{-0.8}$$

$$= 0.953 \text{ (3 s.f.)}$$

(b) If X is number of people with the health problem in a sample of size n, $X \sim B(n, 0.01)$.
If $n > 50$, $X \sim P(0.01n)$ approximately.

$P(X \geqslant 1) = 1 - P(X = 0) = 1 - e^{-0.01n}$
So $P(X \geqslant 1) > 0.95 \Rightarrow 1 - e^{-0.01n} > 0.95$
i.e. $e^{-0.01n} < 0.05$

By trial and improvement, $e^{-2.99} = 0.0502 > 0.05$, $e^{-3} = 0.049 < 0.05$, so $0.01n = 3 \Rightarrow n = 300$. (Log theory could be used here.) The minimum number in the sample should be 300

3 (a) $f(x) = k$, $2 \leqslant x \leqslant 12$
Total probability = 1
$\Rightarrow 10 \times k = 1$, i.e. $k = \frac{1}{10}$
$\therefore\ f(x) = \frac{1}{10}$, $2 \leqslant x \leqslant 12$

(b) By symmetry, $E(X) = \frac{1}{2}(2 + 12) = 7$

$$E(X^2) = \frac{1}{10}\int_2^{12} x^2\,dx = \frac{1}{10}\left[\frac{x^3}{3}\right]_2^{12} = 57\tfrac{1}{3}$$

$Var(X) = 57\tfrac{1}{3} - 7^2 = 8\tfrac{1}{3}$

or $Var(X) = \frac{1}{12}(12 - 2)^2 = 8.333 \ldots$ (formula)

Standard deviation $= \sqrt{8\tfrac{1}{3}} = 2.887$ (3 d.p.)

$P(7 - 2.887 < X < 7 + 2.887)$
$= P(4.113 < X < 9.887)$
$= 0.1 \times (9.887 - 4.113)$
$= 0.58$ (2 d.p.)

4 From the sample: $\bar{x} = \dfrac{\sum x}{n} = \dfrac{255.6}{9} = 28.4$

(a) $\bar{X} \sim N\left(\mu, \dfrac{\sigma^2}{n}\right)$, i.e. $\bar{X} \sim N\left(\mu, \dfrac{4^2}{9}\right)$.

90% confidence limits

$$= \bar{x} \pm 1.645\,\frac{\sigma}{\sqrt{n}}$$

$$= 28.4 \pm 1.645 \times \frac{4}{\sqrt{9}}$$

$$= 28.4 \pm 2.193$$

90% confidence interval

$$= (28.4 - 2.193, 28.4 + 2.193)$$

$$= (26.2, 30.6) \text{ (to 1 d.p.)}$$

(b) (i) $\hat{\sigma}^2 = \dfrac{1}{n-1}\left(\sum x^2 - \dfrac{(\sum x)^2}{n}\right)$

$$= \frac{1}{8}\left(7372.26 - \frac{255.6^2}{9}\right) = 14.1525$$

$\hat{\sigma} = 3.762$ (3 d.p.)

(ii) Since the sample size is small, the $t(8)$ distribution is considered. The critical t values, at the 95% level, are ± 2.306.

95% confidence limits

$$= 28.4 \pm 2.306 \times \frac{3.762}{\sqrt{9}} = 28.4 \pm 2.892$$

Interval $= (25.5, 31.3)$ (to 1 d.p.)

5 For the binomial distribution,
$E(X) = np = 10 \times 0.4 = 4$

$Var(X) = npq = 4 \times 0.6 = 2.4$

By the central limit theorem, since the sample size is large,

$$\bar{X} \sim N\left(4, \frac{2.4}{60}\right), \text{ i.e. } \bar{X} \sim N(4, 0.04)$$

$$P(\bar{X} < 3.5) = P\left(Z < \frac{3.5 - 4}{\sqrt{0.04}}\right)$$

$$= P(Z < -2.5)$$

$$= 0.0062$$

Statistics 2 *(continued)*

6 Let X be the number of trees with the infestation. Assuming that trees are infested independently, with probability p, $X \sim B(n, p)$

$H_0: p = 0.35$

$H_1: p > 0.35$

(a) If $p = 0.35$, $X \sim B(10, 0.35)$

At the 10% level,

H_0 is rejected if $P(X \geqslant 6) < 0.1$.

Using cumulative binomial tables,

$P(X \geqslant 6) = 1 - P(X \leqslant 5)$

$\quad = 1 - 0.9051 = 0.0949$

Since $P(X \geqslant 6) < 0.1$, H_0 is rejected.

There is evidence, at the 10% level, that the percentage of trees that are infested is greater than 35%, indicating that the trees should be felled.

(b) If $p = 0.35$, $X \sim B(30, 0.35)$

n is large such that $np = 10.5 > 5$ and

$nq = 19.5 > 5$ so use normal approximation.

$npq = 10.5 \times 0.65 = 6.825$

so $X \sim N(10.5, 6.825)$ approximately.

At the 10% level, H_0 is rejected if $z > 1.282$.

Test $x = 14$.

Applying a continuity correction,

$$z = \frac{13.5 - 10.5}{\sqrt{6.825}} = 1.148$$

Since $z < 1.282$, H_0 is not rejected.

There is not enough evidence to say that the percentage of infected trees greater than 35%. The trees should be treated with chemicals.

7 (a) $\displaystyle\int_{\text{all } x} f(x)\, dx = 1$

so $\quad 1 = k \displaystyle\int_0^1 (x - x^3)\, dx$

$\quad = k \left[\dfrac{x^2}{2} - \dfrac{x^4}{4} \right]_0^1$

$\quad = k \left(\dfrac{1}{2} - \dfrac{1}{4} \right)$

$\quad = \dfrac{1}{4} k$

$\therefore \quad k = 4$

(b) $\mu = E(X) = \displaystyle\int_0^1 x f(x)\, dx = 4 \int_0^1 x(x - x^3)\, dx$

$\quad = 4 \displaystyle\int_0^1 (x^2 - x^4)\, dx$

$\quad = 4 \left[\dfrac{x^3}{3} - \dfrac{x^5}{5} \right]_0^1$

$\quad = 4 \left(\dfrac{1}{3} - \dfrac{1}{5} \right)$

$\quad = \dfrac{8}{15}$

(c) $P(X < 0.5) = 4 \displaystyle\int_0^{0.5} (x - x^3)\, dx$

$\quad = 4 \left[\dfrac{x^2}{2} - \dfrac{x^4}{4} \right]_0^{0.5}$

$\quad = 0.4375$

(d) $E(X^2) = \displaystyle\int_0^1 x^2 f(x)\, dx = 4 \int_0^1 x^2 (x - x^3)\, dx$

$\quad = 4 \displaystyle\int_0^1 (x^3 - x^5)\, dx$

$\quad = 4 \left[\dfrac{x^4}{4} - \dfrac{x^6}{6} \right]_0^1$

$\quad = \dfrac{1}{3}$

$\text{Var}(X) = E(X^2) - \mu^2$

$\quad = \dfrac{1}{3} - \left(\dfrac{8}{15} \right)^2$

$\quad = 0.0489 \ (3 \text{ s.f.})$

Statistics 2 *(continued)*

8 (a) $F(2) = 1$
$\Rightarrow c(8 - 4 - 1) = 1$ so $c = \frac{1}{3}$

(b) $P(X > 1\frac{1}{2}) = 1 - F(1\frac{1}{2})$
$= 1 - \frac{1}{3}(4 \times 1\frac{1}{2} - (1\frac{1}{2})^2 - 1)$
$= \frac{1}{12}$

(c) For $0 \leqslant x \leqslant 1$,
$$f(x) = \frac{d}{dx}(\tfrac{2}{3}x) = \tfrac{2}{3}$$

For $1 \leqslant x \leqslant 2$,
$$f(x) = \frac{d}{dx}(\tfrac{1}{3}(4x - x^2 - 1))$$
$$= \tfrac{1}{3}(4 - 2x) = \tfrac{2}{3}(2 - x)$$

For $x \leqslant 0$, $x \geqslant 2$, $f(x) = 0$

(d) $E(X) = \int_0^1 \frac{2}{3}x\,dx + \int_1^2 \frac{2}{3}(2x - x^2)dx$
$$= \left[\frac{1}{3}x^2\right]_0^1 + \frac{2}{3}\left[x^2 - \frac{x^3}{3}\right]_1^2$$
$$= \tfrac{1}{3} + \tfrac{2}{3}(4 - \tfrac{8}{3} - (1 - \tfrac{1}{3}))$$
$$= \tfrac{7}{9}$$

9 H_0: Left- or right-handedness and the ability to complete the task are independent.
H_1: There is an association between them.

Expected frequencies

Right-handed, completed task: $\dfrac{150 \times 120}{200} = 90$

	Completed	Not	Totals
Right	90	60	150
Left	30	20	50
Totals	120	80	200

$v = (2 - 1)(2 - 1) = 1$. Consider the $\chi^2(1)$ distribution. From tables, the critical (5%) value is 3.841, so H_0 is rejected if $\chi^2 > 3.841$.

O	E	$\sum \dfrac{(|O - E| - 0.5)^2}{E}$
83	90	0.469 ...
67	60	0.704 ...
37	30	1.408 ...
13	20	2.112 ...
$\sum O = 160$	$\sum E = 160$	4.694 ...

$$\chi^2 = \sum \frac{(|O - E| - 0.5)^2}{E} = 4.694 \text{ (3 d.p.)}$$

Since $\chi^2 > 3.841$, H_0 is rejected. There is evidence of an association between right- or left-handedness and the ability to complete the task in the given time.

Mechanics 2

1 (a) Using $y = x \tan \theta - \dfrac{gx^2}{2v^2}(1 + \tan^2 \theta)$

gives $2.5 = 20 \tan \theta - \dfrac{9.8 \times 20^2}{2 \times 25^2}(1 + \tan^2 \theta)$

$3.136 \tan^2 \theta - 20 \tan \theta + 5.636 = 0$

$\tan \theta = 6.082 \ldots$ or $\tan \theta = 0.2954 \ldots$

$\tan^{-1}(0.2954 \ldots) = 16.5^0$

so a possible angle of projection is 16.5^0.

(b) Vertically: using $s = ut + \frac{1}{2}at^2$ gives

$-0.5 = 25 \sin 16.5^0 t - 4.9t^2$

$4.9t^2 - 25 \sin 16.5^0 t - 0.5 = 0$

$t = 1.516 \ldots$ or $t = -0.067 \ldots$

since $t > 0$

$d = 25 \cos 16.5^0 \times 1.516 - 20$

The distance is 16.3 m to 3 s.f.

2 (a) $x = t^3 - 6t^2 + 3$

$\dot{x} = 3t^2 - 12t = 3t(t - 4)$

When $t = 0$, $\dot{x} = 0$ so initially at rest. At rest again when $t = 4$

$4^3 - 6 \times 4^2 + 3 = -29$

Max distance in negative direction is 29 m.

(b) The particle is at its initial position when

$t^3 - 6t^2 = 0 \Rightarrow t^2(t - 6) = 0$

Particle passes through initial position when $t = 6$.

This gives $\dot{x} = 3 \times 6^2 - 12 \times 6 = 36 \text{ ms}^{-1}$.

3 (a) $\mathbf{r} = (4t^2 + 5)\mathbf{i} + 2t^3 \mathbf{j}$

$\dot{\mathbf{r}} = 8t\mathbf{i} + 6t^2 \mathbf{j}$

When $t = 1$, $\dot{\mathbf{r}} = 8\mathbf{i} + 6\mathbf{j}$ so the speed $= 10 \text{ ms}^{-1}$.

(b) P is moving parallel to

$\mathbf{i} + 3\mathbf{j} \Rightarrow 6t^2 = 24t \Rightarrow 6t(t - 4) = 0$

Since $t > 0$ this gives $t = 4$.

4 (a) 3 cm (by symmetry).

(b) $\bar{y} = \dfrac{24 \times 2 - 4 \times 3}{20} = 1.8 \text{ cm}$

(c)

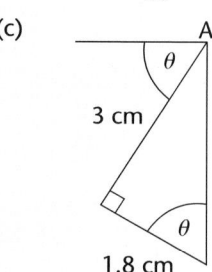

$\theta = \tan^{-1}\left(\frac{3}{1.8}\right) = 59.0^0$

5 (a) Area rectangle : area triangle $= 3 : 1$

so mass of rectangle $= 0.6$ kg

mass of triangle $= 0.2$ kg

$\bar{x} = \dfrac{0.6 \times 7.5 + 0.2 \times 18\frac{1}{3}}{0.8} = 10.2 \text{ cm}$

(b)

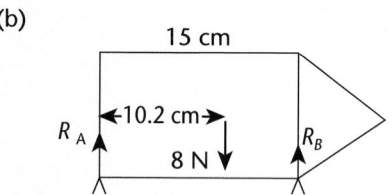

$15R_B = 10.2 \times 8 \Rightarrow R_B = 5.44 \text{ N}$

$R_A = 8 - 5.44 = 2.56 \text{ N}$

Mechanics 2 (continued)

6 (a)

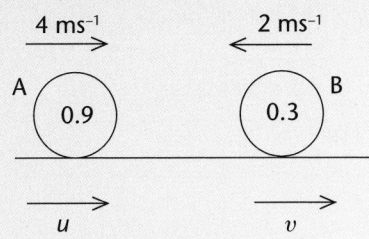

Conservation of momentum:
$0.9 \times 4 - 0.3 \times 2 = 0.9u + 0.3v$

so $3u + v = 10$ (1)

Newton's experimental law:

$v - u = \frac{1}{3} \times 6 = 2$ (2)

From (1) and (2) $4u = 8$
giving $u = 2$, $v = 4$.

The speeds of A and B after the collision are 2 ms^{-1} and 4 ms^{-1} respectively.

(b) Impulse = change in momentum

     $= 0.3 \times 4 - 0.3 \times (-2)$

     $= 1.8$ Ns

(c) Total KE before impact

     $= \frac{1}{2} \times 0.9 \times 4^2 + \frac{1}{2} \times 0.3 \times 2^2 = 7.8$ J

Total KE after impact

     $= \frac{1}{2} \times 0.9 \times 2^2 + \frac{1}{2} \times 0.3 \times 4^2 = 4.2$ J

Loss of mechanical energy = 3.6 J

7

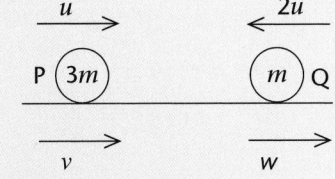

Conservation of momentum:

$3mu - 2mu = 3mv + mw$

giving: $3v + w = u$ (1)

Newton's experimental law:

      $w - v = 3ue$ (2)

(1)–(2) gives $4v = u(1 - 3e)$

so $v = \frac{u}{4}(1 - 3e)$

   $v < 0 \Rightarrow (1 - 3e) < 0$

      $\Rightarrow e > \frac{1}{3}$

8

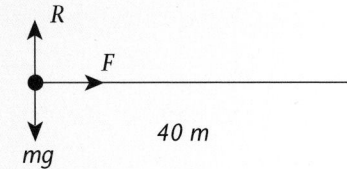

vertically: $R = mg$

horizontally: $F = \dfrac{m \times 15^2}{40}$

$\mu = \dfrac{F}{R} = \dfrac{15^2}{40 \times 9.8} = 0.57$

9

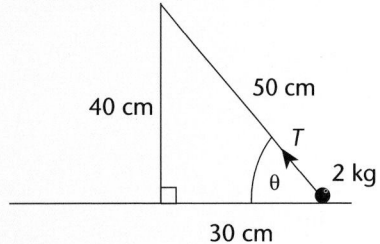

(a)   magnitude of acceleration $= \omega^2 r$

                      $= 3^2 \times 0.3$

                      $= 2.7$ ms^{-2}

(b)   Using $F = ma$: $T \cos\theta = 2 \times 2.7$

      so: $T \times \frac{3}{5} = 5.4$

      giving: $T = 9$ N

(c)

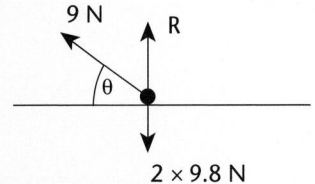

Vertically: $R + 9 \sin\theta = 2 \times 9.8$

              $R = 2 \times 9.8 - 9 \times 0.8 = 12.4$

              $R = 12.4$ N

Mechanics 2 *(continued)*

10 (a) The tension is perpendicular to the direction of motion.

(b) The acceleration is not constant.

(c) Maximum speed occurs at lowest point.

Loss of PE = gain in KE

$$mgl = \tfrac{1}{2}mv^2$$

so, maximum speed is given by $v = \sqrt{2gl}$

11 (a)

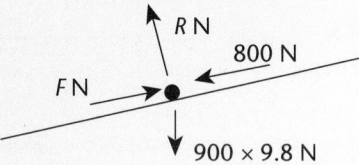

Resolving parallel to the plane:

$$F = 900 \times 9.8 \times \tfrac{1}{15} + 800$$
$$= 1388$$

The driving force is 1388 N

(b) $P = Fv = 1388 \times 20 = 27760$

Power $= 27.76$ kW

(c) Using $F = ma$:

gives: $1388 - 800 = 900a \Rightarrow a = 0.65\ ms^{-2}$.

12 Momentum of ball before impact

$$= 0.2 \times 20 = 4\ \text{Ns}$$

Momentum of ball after impact

$$= 0.2 \times 5 = 1\ \text{Ns}$$

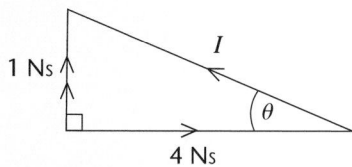

The magnitude of the impulse is given by:

$$I^2 = 1^2 + 4^2 = 17 \Rightarrow I = \sqrt{17}$$

$$\tan \theta = \tfrac{1}{4} \Rightarrow \theta = 14.0^0$$

The impulse has magnitude $\sqrt{17}$ Ns at an angle of 14.0^0 above the horizontal.

13 (a) The velocity at time t is given by:
$$\dot{x} = 12t^2 - 6t = 6t(2t - 1)$$

(b) $\dot{x} = 0$ when $t = 0$ or $t = 0.5$ so the particle changes direction when $t = 0.5$.

$t = 0 \Rightarrow x = 7$

$t = 0.5 \Rightarrow x = 6.75$

The distance travelled is 0.25 m.

(c) P is at its starting point when
$$4t^3 - 3t^2 = 0$$
$$\Rightarrow t^2(4t - 3) = 0$$
$$\Rightarrow t = 0,\ \text{or}\ t = 0.75$$

P returns to its start point after 0.75 s

Mechanics 2 *(continued)*

14 Using $F = ma$: $-kv^2 = m\dfrac{dv}{dt}$

$\dfrac{dv}{dt} = -3$ when $v = 20$

so: $-\dfrac{k}{m} = \dfrac{1}{400} \times -3$

$\Rightarrow \dfrac{dv}{dt} = -\dfrac{3}{400}v^2$

$\Rightarrow \displaystyle\int v^{-2}dv = \int -\dfrac{3}{400}\,dt$

giving: $-\dfrac{1}{v} = -\dfrac{3t}{400} + C$

when $t = 0$, $v = 20$

so: $-\dfrac{1}{20} = C$

This gives $\dfrac{1}{v} = \dfrac{3t}{400} + \dfrac{1}{20} = \dfrac{3t + 20}{400}$

$\Rightarrow v = \dfrac{400}{3t + 20}$ as required.

When $t = 0$: KE $= \frac{1}{2} \times 2 \times 20^2 = 400$ J

When $t = 4$: KE $= \frac{1}{2} \times 2 \times 12.5^2 = 156.25$ J

Work done by retarding force has magnitude
$400 - 156.25 = 243.75$ J.

15 (a) At the lowest point, the loss of gravitational potential energy has been converted to elastic potential energy.

Using $mgh = \dfrac{\lambda x^2}{2l}$

gives $mg(2 + x) = \dfrac{5mgx^2}{4}$

$\Rightarrow 5x^2 - 4x - 8 = 0$.

$\Rightarrow x = 1.73$ or $x = -0.927$ to 3 s.f.

When the particle is brought to rest, the extension is 1.73 m so the object is 3.73 m below O.

(b) Using $T = \dfrac{\lambda}{l}x$

gives $T = \dfrac{5mg}{2} \times 1.73$

Using $F = ma$ gives:

$\dfrac{5mg}{2} \times 1.73 - mg = ma$

Taking $g = 9.8$ gives:

$a = 2.5 \times 9.8 \times 1.73 - 9.8$

$= 32.585$

The object has acceleration 32.6 ms^{-2} to 3 s.f. at its lowest point.

Mechanics 2 *(continued)*

16 (a)

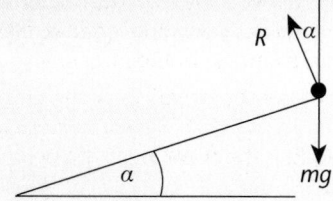

Vertically: $R \cos \alpha = mg$ (1)

Horizontally: $R \sin \alpha = \dfrac{m \times 20^2}{80}$ (2)

(2) ÷ (1) gives $\tan \alpha = \dfrac{20^2}{80 \times 9.8}$

$$\Rightarrow \alpha = 27.0^0$$

(b)

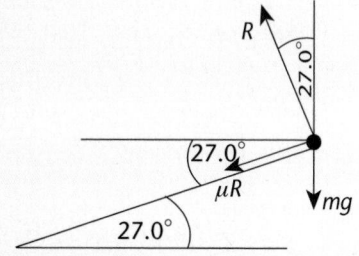

vert: $R \cos 27^0 = mg + \mu R \sin 27^0$ (3)

horiz: $R \sin 27^0 + \mu R \cos 27^0 = \dfrac{m \times 28^2}{80}$ (4)

(4) ÷ (3) gives:

$$\frac{\sin 27^0 + \mu \cos 27^0}{\cos 27^0 - \mu \sin 27^0} = \frac{28^2}{80 \times 9.8}$$

rearranging gives:

$$\mu = \frac{28^2 \cos 27^0 - 80 \times 9.8 \sin 27^0}{80 \times 9.8 \cos 27^0 + 28^2 \sin 27^0}$$

$$\Rightarrow \mu = 0.325 \text{ to 3 s.f.}$$

17 (a)   Horizontally:

$$T \sin 30^0 = \frac{0.2 \times 1.2^2}{0.4}$$

$$\Rightarrow T = 1.44$$

The tension is 1.44 N.

Vertically:

$$1.44 \cos 30^0 + R = 0.2 \times 9.8$$

$$\Rightarrow R = 0.713$$

The reaction is 0.713 N to 3 s.f.

(b)   Vertically: the new tension is given by:

$$T \cos 30^0 = 0.2 \times 9.8 \Rightarrow T = 2.263$$

Horizontally: $2.263 \sin 30^0 = 0.2\omega^2 \times 0.4$

giving: $\omega = 3.76$

The minimum angular speed is 3.76 rad sec^{-1}.

18 (a)   By the conservation of energy principle:
KE gained = PE lost

giving: $\frac{1}{2} \times 0.1 \times v^2 = 0.1g \times 0.6(1 - \cos \theta)$

$$\Rightarrow v^2 = 1.2g(1 - \cos \theta)$$

(b)   Resolving along BC:

$$0.1g \cos \theta - R = \frac{0.1 \times 1.2g(1 - \cos \theta)}{0.6}$$

rearranging gives:

$$R = g(0.3 \cos\theta - 0.2)$$

(c) $R = 0 \Rightarrow 0.3 \cos \theta = 0.2$

$$\Rightarrow \cos\theta = \tfrac{2}{3}$$

$$\Rightarrow \theta = 48.2^0$$

Decision Mathematics 2

1 Assume A uses Strategy W with probability p and strategy X with probability $(1 - p)$.

Expected gain for A is:

$4p + 6(1 - p)$ if B uses strategy Y.

$8p + 3(1 - p)$ if B uses strategy Z.

Optimal when

$$4p + 6(1 - p) = 8p + 3(1 - p)$$

$$\Rightarrow -2p + 6 = 5p + 3$$

$$\Rightarrow p = \tfrac{3}{7}$$

Assume B uses strategy Y with probability q and strategy Z with probability $(1 - q)$.

Expected loss for B is:

$4q + 8(1 - q)$ if A uses strategy W

$6q + 3(1 - q)$ if A uses strategy X.

Optimal when

$$4q + 8(1 - q) = 6q + 3(1 - q)$$

$$\Rightarrow -4q + 8 = 3q + 3$$

$$\Rightarrow q = \tfrac{5}{7}$$

The optimal strategy for A is to use strategy W with probability $\tfrac{3}{7}$ and

strategy X with probability $\tfrac{4}{7}$.

The optimal strategy for B is to use strategy Y with probability $\tfrac{5}{7}$ and

strategy Z with probability $\tfrac{2}{7}$.

The value of the game is $5\tfrac{1}{7}$.

2 (a) The minimum row values are –4, –3 and –6. The maximum of these is –3 showing that the play-safe strategy for Richard is strategy B.

The minimum of the negatives of the column values are –4, –6 and –8. The maximum of these is –4 showing that the play-safe strategy for Judy is strategy X.

(b) $-3 + (-4) \neq 0$ so there is no saddle point.

(c) Comparing the row values shows that Richard's pay-offs are *always* less with strategy C than with strategy A. It follows that strategy C will not be a part of his mixed strategy.

(d) Assume Richard uses strategy A with probability p and strategy B with probability $(1 - p)$.

The expected gains for Richard based on the strategy used by Judy are:

X: $4p - 3(1 - p) = 7p - 3$

Y: $3p + 6(1 - p) = -3p + 6$

Z: $-4p + 8(1 - p) = -12p + 8$

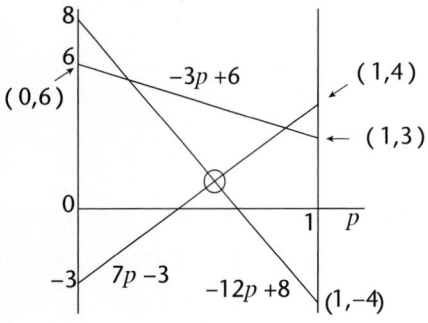

Optimal when $7p - 3 = -12p + 8$

$$\Rightarrow p = \tfrac{11}{19}$$

Richard's optimal strategy is to use strategy A with probability $\tfrac{11}{19}$ and strategy B with probability $\tfrac{8}{19}$.

Decision Mathematics 2 (continued)

3 (a) The objective function is $4a + 9b + 12c$.

(b) The formulation of the task as a linear programming problem is:

Maximise: $P = 4a + 9b + 12c$

Subject to: $6a + 15b + 20c \leqslant 100$

$a + b + c \leqslant 10$

and $a \geqslant 0, b \geqslant 0, c \geqslant 0$.

(c) In the solution a, b and c must be integers.

4 (a)

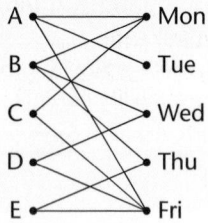

(b)

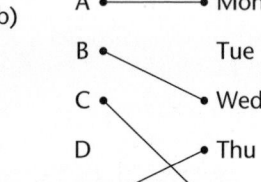

(c)

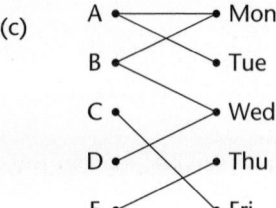

The alternating path is:

D – Wed – B – Mon – A – Tue.

The new matching is:

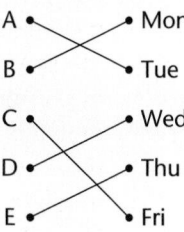

5 (a) The initial tableau is:

P	x	y	z	s	t	val
1	−1	−2	−3	0	0	0
0	2	6	1	1	0	10
0	1	4	5	0	1	16

(b) The first iteration produces:

P	x	y	z	s	t	val
1	$-\frac{2}{5}$	$\frac{2}{5}$	0	0	$\frac{3}{5}$	$9\frac{3}{5}$
0	$\frac{9}{5}$	$\frac{26}{5}$	0	1	$-\frac{1}{5}$	$6\frac{4}{5}$
0	$\frac{1}{5}$	$\frac{4}{5}$	1	0	$\frac{1}{5}$	$3\frac{1}{5}$

The second iteration produces:

P	x	y	z	s	t	val
1	0	$\frac{14}{9}$	0	$\frac{2}{9}$	$\frac{5}{9}$	$11\frac{1}{9}$
0	1	$\frac{26}{9}$	0	$\frac{5}{9}$	$-\frac{1}{9}$	$3\frac{7}{9}$
0	0	$\frac{2}{9}$	1	$-\frac{1}{9}$	$\frac{2}{9}$	$2\frac{4}{9}$

The maximum value of P is $11\frac{1}{9}$.

This occurs when $x = 3\frac{7}{9}$, $y = 0$, $z = 2\frac{4}{9}$.

(c) This represents the optimum solution because there are no negative values in the objective row of the tableau.

Index